Data Mining for Geoinformatics

Guido Cervone • Jessica Lin • Nigel Waters

Editors

Data Mining for Geoinformatics

Methods and Applications

 Springer

Editors
Guido Cervone
Department of Geography and Institute
    for CyberScience
The Pennsylvania State University
State College, PA, USA

Research Application Laboratory
National Center for Atmospheric Research
Boulder, CO, USA

Nigel Waters
Center of Excellence in GIS
George Mason University
Fairfax, VA, USA

Jessica Lin
Department of Computer Science
George Mason University
Fairfax, VA, USA

ISBN 978-1-4899-8574-3   ISBN 978-1-4614-7669-6 (eBook)
DOI 10.1007/978-1-4614-7669-6
Springer New York Heidelberg Dordrecht London

Printed on acid-free paper

Springer is part of Springer Science+Business Media (www.springer.com)

Introduction

In March 1999, the National Center for Geographic Information and Analysis based at the University of California at Santa Barbara held a workshop on Discovering Geographic Knowledge in Data-Rich Environments. This workshop resulted in a seminal, landmark, edited volume (Miller and Han 2001a) that brought together research papers contributed by many of the participants at that workshop. In their introductory essay, Miller and Han (2001b) observe that geographic knowledge discovery (GKD) is a nontrivial, special case of knowledge discovery from databases (KDD). They note that this is in part due to the distinctiveness of geographic measurement frameworks, problems incurred and resulting from spatial dependency and heterogeneity, the complexity of spatiotemporal objects and rules, and the diversity of geographic data types. Miller and Han's book was enormously influential and, since publication, has garnered almost 350 citations. Not only has it been well cited but in 2009 a second edition was published. Our current volume revisits many of the themes introduced in Miller and Han's book. In the collection of six papers presented here, we address current concerns and developments related to spatiotemporal data mining issues in remotely sensed data, problems in meteorological data such as tornado formation, simulations of traffic data using OpenStreetMap, real-time traffic applications of data stream mining, visual analytics of traffic and weather data, and the exploratory visualization of collective, mobile objects such as the flocking behavior of wild chickens.

Our volume begins with a discussion of computation in hyperspectral imagery data analysis by Mark Salvador and Ron Resmini. Hyperspectral remote sensing is the simultaneous acquisition of hundreds of narrowband images across large regions of the electromagnetic spectrum. Hyperspectral imagery (HSI) contains information describing the electromagnetic spectrum of each pixel in the scene, which is also known as the spectral signature. Although individual spectral signatures are recognizable, knowable, and interpretable, algorithms with a broad range of sophistication and complexity are required to sift through the immense quantity of spectral signatures and to extract information leading to the formation of useful products. Large hyperspectral data cubes were once thought to be a significant data mining and data processing challenge, prompting research in algorithms, phenomenology, and computational methods to speed up analysis.

Although modern computer architectures make quick work of individual hyperspectral data cubes, the preponderance of data increases significantly year after year. HSI analysis still relies on accurate interpretation of both the analysis methods and the results. The discussion in this chapter provides an overview of the methods, algorithms, and computational techniques for analyzing hyperspectral data. It includes a general approach to analyzing data, expands into computational scope, and suggests future directions.

The second chapter, authored by Amy McGovern, Derek H. Rosendahl, and Rodger Brown, uses time series data mining techniques to explain tornado genesis and development. The mining of time series data has gained a lot of attention from researchers in the past two decades. Apart from the obvious problem of handling the typically large size of time series databases—gigabytes or terabytes are not uncommon—most classic data mining algorithms do not perform or scale well on time series data. This is mainly due to the inherent structure of the data, that is, high dimensionality and feature correlation, which pose challenges that render classic data mining algorithms ineffective and inefficient. Besides individual time series, it is also common to encounter time series with one or more spatial dimensions. These spatiotemporal data can appear in the form of spatial time series or moving object trajectories. Existing data mining techniques offer limited applicability to most commercially important and/or scientifically challenging spatiotemporal mining problems, as the spatial dimensions add an increased complexity to the analysis of the data. To manipulate the data efficiently and discover nontrivial spatial, temporal, and spatiotemporal patterns, there is a need for novel algorithms that are capable of dealing with the challenges and difficulties posed by the temporal aspect of the data (time series) as well as handling the added complexity due to the spatial dimensions.

The mining of spatiotemporal data is particularly crucial for fields such as the earth sciences, as its success could lead to significant scientific discovery. One important application area for spatiotemporal data mining is the study of natural phenomena or hazards such as tornadoes. The forecasting of tornadoes remains highly unreliable – its high false alarm rate causes the public to disregard valid warnings. There is clearly a need for scientists to explore ways to understand environmental factors that lead to tornado formations. Toward that end, the authors of this chapter propose novel spatiotemporal algorithms that identify rules, salient variables, or patterns predictive of tornado formation. Their approach extends existing algorithms that discover repetitive patterns called time series motifs. The multidimensional motifs identified by their algorithm can then be used to learn predictive rules. In their study, they identify ten statistically significant attributes associated with tornado formation.

In the third chapter, Guido Cervone and Pasquale Franzese discuss the estimation of the release rate for the nuclear accident that occurred at the Fukushima Daiichi nuclear power plant. Unlike a traditional source detection problem where the location of the source is one of the unknowns, for this accident the main task is to determine the amount of radiation leaked as a function of time. Determining the amount of radiation leaked as a result of the accident is of paramount importance to understand the extent of the disaster and to improve the safety of existing and future nuclear power plants.

A new methodology is presented that uses spatiotemporal data mining to reconstruct the unsteady release rate using numerical transport and dispersion simulations together with ground measurements distributed across Japan. As in the previous chapter, the time series analysis of geographically distributed data is the main scientific challenge. The results show how geoinformatics algorithms can be used effectively to solve this class of problems.

Jorg Dallmeyer, Andreas Lattner, and Ingo Timm, the authors of the fourth chapter, explain how to build a traffic simulation using OpenStreetMap (OSM), perhaps the best known example of a volunteered geographic database that relies on the principles of crowd sourcing. Their chapter begins with an overview of their methodology and then continues with a discussion of the characteristics of the OSM project. While acknowledging the variable quality of the OSM network, the authors demonstrate that it is normally sufficient for the traffic simulation purposes. OSM uses an XML format, and they suggest that it is preferable to parse this for input to a Geographic Information System (GIS). Their process involves the use of an SAX (Simple API for XML) parser and subsequently the open source GIS toolkit GeoTools. This toolkit is also used to generate the initial graph of the road network. Additional processing steps are then necessary to generate important real-world components of the road network, including traffic circles, road type and road user information, and bus routes among other critical details that are important for creating realistic and useful traffic simulations.

A variety of simulation models that focus on multimodal traffic in urban scenarios are produced. The various modes include passenger cars, trucks, buses, bicycles, and pedestrians. The first of these is a space-continuous simulation based on the Nagel-Schreckenberg model (NSM). The bicycle model is a particularly interesting contribution of this chapter since, as the authors correctly observe, it has been little studied in transportation science so far. Similarly pedestrians too have been largely neglected, and integrating both bicycles and pedestrians into the traffic simulation is a noteworthy contribution. An especially intriguing aspect of the research by Dallmeyer and his colleagues is the section of their chapter that describes learning behavior in the various traffic scenarios. Supervised, unsupervised, and reinforcement learning are all examined. In the former, the desired output of the learning process is known in advance. This is not the case in the latter two instances. In addition, in reinforcement learning, the driver, cyclist, or pedestrian receives no *direct* feedback.

The final section of this chapter considers a series of case studies based on Frankfurt am Main, Germany. The simulations based on this city are shown to be able to predict traffic jams with a greater than 80% success rate. Subsequent research will focus on models to predict gas consumption and CO_2 emissions.

The work by Sandra Geisler and Christoph Quix, the authors of our fifth chapter, relies, in part, on traffic simulations similar to those discussed by Dallmeyer and his colleagues. This chapter describes a complete system for analyzing the large data sets that are generated in intelligent transportation systems (ITS) from sensors that are now being integrated into car monitoring systems. Such sensor systems are designed to increase both comfort and, more importantly, safety. The safety component that involves warning surrounding vehicles of, for example, a sudden braking action has been termed Geocasting or GeoMessaging. The goal of ITS is to monitor the state of the traffic over large areas at the lowest possible costs. In order to produce an effective transportation management system using these data, Geisler and Quix observe that they must handle extremely large amounts of data, in real time with high levels of accuracy. The aim of their research is to provide a framework for evaluating data stream ITS using various data mining procedures. This framework incorporates traffic simulation software, a Data Stream Management System (DSMS), and data stream mining algorithms for mining the data stream. In addition, the Massive Online Analysis (MOA) framework that they exploit permits flexibility in monitoring data quality using an ontology-based approach. A mobile Car-to-X (C2X) communication system is integrated into the structure as part of the communication system. The architecture of the system was initially designed as part of the CoCar Project. The system ingests data from several primary sources: cooperative cars, floating phone data, and stationary sources. The DSMS includes aggregation and integration steps that are followed by data accuracy assessments and utilizes the Global Sensor Network system. Following this, data mining algorithms are used for queue end detection and traffic state analysis. Historical and spatial data are imported prior to the export of the traffic messaging. The spatial database resolves the transportation network into 100 m arcs. To determine the viability of the system, data are generated using the VISSIM traffic simulation software. A particularly significant feature of the authors' approach is to use a flexible set of data quality metrics in the DSMS. These metrics are application, content, and query specific.

The effectiveness of the framework is examined in a series of case studies. The first set of case studies concerned traffic queue end detection based on the detection of hazards resulting from traffic congestion. A second group of studies used a road network near Dusseldorf, Germany, and involved traffic state estimation based on four states: free, dense, slow moving, and congested. The chapter concludes with a discussion of other ways in which data streaming management systems could be applied to ITS problems, including the simulation of entire days of traffic with high variance conditions that would include both bursts of congestion and relatively calm interludes.

Snow removal and the maintenance of safe driving conditions are perennial concerns for many high-latitude cities in the northern hemisphere during the winter months. Our sixth chapter by Yuzuru Tanaka and his colleagues, Jonas Sjöbergh, Pavel Moiseets, Micke Kuwahara, Hajime Imura, and Tetsuya Yoshida, at the University of Hokkaido, in Sapporo, Japan, develops a variety of software and data mining tools within a federated environment for addressing and resolving these predicaments. Although snow removal presents operational difficulties for many cities, few face the challenges encountered in Sapporo where the combination of a population of almost two million and an exceptionally heavy snowfall makes timely and efficient removal an ongoing necessity to avoid unacceptable levels of traffic congestion. Data mining techniques use data from taxis and so-called probe cars, another form of volunteered geographic information, to track vehicle location and speed. In addition, these data are supplemented with meteorological sensor and snow removal data along with claims to call centers and social media data from Twitter.

The chapter proposes and develops an integrated geospatial visualization and analytics environment. The enabling, integration technology is the Webble World environment developed at Tanaka's Meme Media Laboratory at the University of Hokkaido. The visual components of this environment, known as Webbles, are then integrated into federated applications. To integrate the various components of this system, including the GIS, statistical and knowledge discovery tools, and social networking systems (SNS) such as Twitter, specific wrappers are written for Esri's ArcView software and generic wrappers are developed in R and Octave for the remaining components. The chapter provides a detailed description of the Webble World framework as well as information on how readers may access the system and experiment for themselves.

Case studies for snowfall during 2010 and 2011 are described when data for about 2,000 taxis were accessed. The data are processed into street segments for the Sapporo road network. The street segments are then grouped together using a spherical k-means clustering algorithm. Differences in traffic characteristics, for example, speed, congestion, and other attributes, between snowfall and non-snowfall and before and after snow removal are then visualized. The beauty of the system is the ease with which the Webble World environment integrates the various newly federated data streams. In addition, mash-ups of the probe car and the weather station, call center complaints, and Twitter tweets are also discussed.

Chapter 7, our final chapter, written by Tetsuo Kobayashi and Harvey Miller, concerns exploratory spatial data analysis for the visualization of collective mobile objects data. Recent advances in mobile technology have produced a vast amount of spatiotemporal trajectory data from moving objects. Early research work on moving objects has focused on techniques that allow efficient storage and querying of data. In recent years, there has been an increasing interest in finding patterns, trends, and relationships from moving object trajectories. In this chapter, the authors introduce a visualization system that summarizes (aggregates) moving objects based on their spatial similarity, using different levels of temporal granularity. Its ability to process a large amount of data and produce a compact representation of these data allows the detection of interesting patterns in an efficient manner. In addition, the user-interactive capability facilitates dynamic visual exploration and a deep understanding of data. A case study on wild chicken movement trajectories shows that the combination of spatial aggregation and varying temporal granularity is indeed effective in detecting complex flocking behavior.

Washington D.C.,   Guido Cervone, Jessica Lin
                                                                Nigel Waters

Miller HJ, Han J (2001a, 2009) Geographic data mining and knowledge discovery. Taylor and Francis, London

Miller HJ, Han J (2001b) Geographic data mining and knowledge discovery: an overview. Ch 1, pp 3–32, in Miller and Han, op. cit

Contents

Computation in Hyperspectral Imagery (HSI) Data Analysis: Role and Opportunities

Mark Salvador and Ron Resmini

Abstract Successful quantitative information extraction and the generation of useful products from hyperspectral imagery (HSI) require the use of computers. Though HSI data sets are stacks of images and may be viewed as images by analysts, harnessing the full power of HSI requires working primarily in the spectral domain. Algorithms with a broad range of sophistication and complexity are required to sift through the immense quantity of spectral signatures comprising even a single modestly sized HSI data set. The discussion in this chapter will focus on the analysis process that generally applies to all HSI data and discuss the methods, approaches, and computational issues associated with analyzing hyperspectral imagery data.

Keywords Remote sensing • Hyperspectral • Hyperspectral imagery • Multi-spectral • VNIR/SWIR • LWIR • Computational science

1 Introduction

Successful quantitative information extraction and the generation of useful products from hyperspectral imagery (HSI) require the use of computers. Though HSI data sets are stacks of images and may be viewed as images by analysts ('literal' analysis), harnessing the full power of HSI requires working primarily in the spectral domain. And though individual spectral signatures are recognizable, knowable, and

M. Salvador (✉)
Integrated Sensing and Information Systems, Exelis Inc., 12930 Worldgate Drive, Herndon, VA 20170, USA
e-mail: mzsalvador@icloud.com

R. Resmini
The MITRE Corporation, 7515 Colshire Drive, McLean, VA 22102, USA
e-mail: rresmini@mitre.org

G. Cervone et al. (eds.), *Data Mining for Geoinformatics: Methods and Applications*, DOI 10.1007/978-1-4614-7669-6_1, © Springer Science+Business Media New York 2014

interpretable,[1] algorithms with a broad range of sophistication and complexity are required to sift through the immense quantity of spectral signatures comprising even a single modestly sized HSI data set and to extract information leading to the formation of useful products ('nonliteral' analysis).

But first, what is HSI and why acquire and use it? Hyperspectral remote sensing is the collection of hundreds of images of a scene over a wide range of wavelengths in the visible ($\sim$0.40 micrometers or μm) to longwave infrared (LWIR, $\sim$14.0 μm) region of the electromagnetic spectrum. Each image or band samples a small wavelength interval. The images are acquired simultaneously and are thus coregistered with one another forming a stack or image cube. The majority of hyperspectral images (HSI) are from regions of the spectrum that are outside the range of human vision which is $\sim$0.40 to $\sim$0.70 μm. Each HSI image results from the interaction of photons of light with matter: materials reflect (or scatter), absorb, and/or transmit electromagnetic radiation (see, e.g., Hecht 1987; Hapke 1993; Solé et al. 2005; Schaepman-Strub et al. 2006; and Eismann 2012, for detailed discussions of these topics fundamental to HSI). Absorbed energy is later emitted (and at longer wavelengths—as, e.g., thermal emission). The light energy which is received by the sensor forms the imagery. Highly reflecting materials form bright objects in a band or image; absorbing materials (from which less light is reflected) form darker image patches. Ultimately, HSI sensors detect the radiation reflected (or scattered) from objects and materials; those materials that mostly absorb light (and appear dark) are also reflecting (or scattering) some photons back to the sensor. Most HSI sensors are passive; they only record reflected (or scattered) photons of sunlight or photons self-emitted by the materials in a scene; they do not provide their own illumination as is done by, e.g., lidar or radar systems. HSI is an extension of multispectral imagery remote sensing (MSI; see, e.g., Jensen 2007; Campbell 2007; Landgrebe 2003; Richards and Jia 1999). MSI is the collection of tens of bands of the electromagnetic spectrum. Individual MSI bands or images sample the spectrum over larger wavelength intervals than do individual HSI images.

The discussion in this chapter will focus on the analysis process beginning with the best possible calibrated at-aperture radiance data. Collection managers/data consumers/end users are advised to be cognizant of the various figures of merit (FOM) that attempt to provide some measure of data quality; e.g., noise equivalent spectral radiance (NESR), noise equivalent change of reflectance (NE$\Delta\rho$), noise equivalent change of temperature (NEΔT), and noise equivalent change of emissivity (NE$\Delta\varepsilon$).

What we will discuss generally applies, at some level, to all HSI data: visible/near-infrared (VNIR) through LWIR. There are procedures that are applied to the midwave infrared (MWIR) and LWIR[2] that are not applied to VNIR/shortwave

---

[1]The analyst is encouraged to study and become familiar with several spectral signatures likely to be found in just about every earth remote sensing data set: vegetation, soils, water, concrete, asphalt, iron oxide (rust), limestone, gypsum, snow, paints, fabrics, etc.

[2]The MWIR and LWIR (together or individually) may be referred to as the thermal infrared or TIR.

infrared (SWIR); e.g., temperature/emissivity separation (TES). Atmospheric compensation (AC) for thermal infrared (TIR) spectral image data is different (and, for the MWIR,[3] arguably more complicated) than for the VNIR/SWIR. But such differences notwithstanding, the bulk of the information extraction algorithms and methods (e.g., material detection and identification; material mapping)— particularly after AC—apply across the full spectral range from 0.4 μm (signifying the lower end of the visible) to 14 μm (signifying the upper end of the LWIR).

What we *won't* discuss (and which require computational resources): all the processes that get the data to the best possible calibrated at-aperture radiance; optical distortion correction (e.g., spectral smile); bad/intermittent pixel correction; saturated pixel(s) masking; "NaN" pixel value masking; etc.

Also, we will not rehash the derivation of algorithm equations; we'll provide the equations, a description of the terms, brief descriptions that will give the needed context for the scope of this chapter, and one or more references in which the reader will find significantly more detail.

2 Computation for HSI Data Analysis

2.1 *The Only Way to Achieve Success in HSI Data Analysis*

No amount of computational resources can substitute for practical knowledge of the remote sensing scenario (or problem) for which spectral image (i.e., HSI) data have been acquired. Successful HSI analysis and exploitation are based on the application of several specialized algorithms deeply informed by a detailed understanding of the physical, chemical, and radiative transfer (RT) processes of the scenario for which the imaging spectroscopy data are acquired. Thus, the astute remote sensing data analyst will seek the input of a subject matter expert (SME) knowledgeable of the materials, objects, and events captured in the HSI data. The analyst, culling as many remote sensing and geospatial data sources as possible (e.g., other forms of remote sensing imagery; digital elevation data) should work collaboratively with the SME (who is also culling as many subject matter information sources as possible) through much of the remote sensing exploitation flow—each informing the other about analysis strategies, topics for additional research, and materials/objects/events to be searched for in the data. It behooves the analyst to be a SME; remote sensing is, after all, a tool; one of many today's multi-disciplinary professional should bring to bear on a problem or a question of scientific, technical, or engineering interest.

It s important to state again, no amount of computational resources can substitute for practical knowledge of the problem and its setting for which HSI data have been

---

[3] We will no longer mention the MWIR; though the SEBASS sensor (Hackwell et al. 1996) provides MWIR data, very little has been made available. MWIR HSI is an area for research, however. MWIR data acquired during the day time have a reflective and an emissive component which introduces some interesting complexity for AC.

Fig. 1 The general HSI data
analysis flow. Our discussion
will begin at the box indicated
by *small arrow* in top box:
'Look At/Inspect the Data'

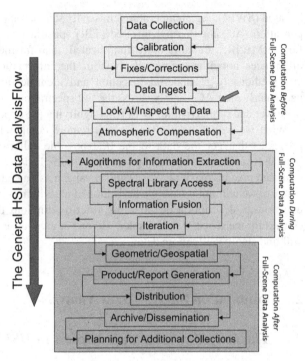

acquired. Even with today's desktop computational resources such as multi-core
central processing units (CPUs) and graphics processing units (GPUs), brute force
attempts to process HSI data without specific subject matter expertise simply lead
to poor results faster. Stated alternatively, computational resources should never be
considered a substitute (or proxy) for subject matter expertise. With these caveats in
mind, let's now proceed to discussing the role of computation in HSI data analysis
and exploitation.

2.2 When Computation Is Needed

The General HSI Data Analysis Flow

The general HSI data analysis flow is shown in Fig. 1. We will begin our discussion
with 'Look At/Inspect the Data' (indicated by small arrow in top box). The flow
chart from this box downwards is essentially the outline for the bulk of this chapter.
The flow reflects the data analyst's perspective though he/she, as a data end-user,
will begin at 'Data Ingest' (again assuming one starts with the best possible, highest-
quality, calibrated at-aperture radiance data).

Though we'll follow the flow of Fig. 1, there is, implicitly, a higher-level clustering of the steps in the figure. This shown by the gray boxes subsuming one or more of the steps and which also form a top-down flow; they more succinctly indicate when computational resources are brought to bear in HSI data analysis. For example, 'Data Ingest', 'Look At/Inspect the Data' and 'Atmospheric Compensation' may perhaps logically fall into something labeled 'Computation *Before* Full-Scene Data Analysis'. An example of this is the use of stepwise least-squares regression analysis[4] to select the best bands and/or combination of bands that best map one or more ground-truth parameters such as foliar chemistry derived by field sampling (and laboratory analysis) at the same time an HSI sensor was collecting (Kokaly and Clark 1999). We refer to this as 'regression remote sensing'; there is a computational burden for the statistical analyses that generate coefficients for one (or more) equations which will then be applied to the remotely sensed HSI data set. The need for computational resources can vary widely in this phase of analysis. The entire pantheon of existing (and steady stream of new) multivariate analysis, optimization, etc., techniques for fitting and for band/band combinations selection may be utilized.

Atmospheric compensation (AC) is another example. There are numerous AC techniques that ultimately require the generation of look-up-tables (LUTs) with RT (radiative transfer) modeling. The RT models are generally tuned to the specifics of the data for which the LUTs will be applied (e.g., sensor altitude, time of day, latitude, longitude, expected ground cover materials); the LUTs may be generated prior to (or at the very beginning of) HSI data analysis.

The second gray box subsumes 'Algorithms for Information Extraction' and all subsequent boxes down to (and including) 'Iteration' (which isn't really a process but a reminder that information extraction techniques should be applied numerous times with different settings, with different spatial and spectral subsets, with in-scene and with library signatures, different endmember/basis vector sets, etc.). This box is labeled 'Computation *During* Full-Scene Data Analysis'.

The third box covers the remaining steps in the flow and is labeled 'Computation *After* Full-Scene Data Analysis'. We won't have much to say about this phase of HSI analysis beyond a few statements about the need for computational resources for geometric/orthorectification post-processing of HSI-derived results and products.

Experienced HSI practitioners may find fault with the admittedly coarse two-tier flow categorization described above. And indeed, they'd have grounds for argument. For example, a PCA may rightly fall into the first gray box 'Computation *Before* Full-Scene Data Analysis'. Calculation of second order statistics for a data cube (see below) and the subsequent generation of a PC-transformed cube for use in data inspection may be accomplished early on (and automatically) in the data analysis process—and not in the middle gray box in Fig. 1. Another example is AC. AC is significantly more than the early-on generation of LUTs. There is the actual process

---

[4]Or principal components regression (PCR) or partial least squares regression (PLSR; see, e.g., Feilhauer et al. 2010).

of applying the LUT with an RT expression to the spectra comprising the HSI cube. This processing (requiring band depth mapping, LUT searching, optimization, etc.) is part of the core HSI analysis process and is not merely a 'simple' LUT-generation process executed early on. Other AC tools bring to bear different procedures that may also look more like 'Computation *During* Full-Scene Data Analysis' such as finding the scene endmembers (e.g., the QUAC tool; see below).

Nonetheless, a structure is needed to organize our presentation and what's been outlined above will suffice. We will thus continue our discussion guided by the diagram in Fig. 1. Exemplar algorithms and techniques for each process will be discussed. Ground rules. (1) Acronyms will be used in the interest of space; an acronym table is provided in an appendix. (2) We will only discuss widely recognized, 'mainstream' algorithms and tools that have been discussed in the literature and are widely used. References are provided for the reader to find out more about any given algorithm or tool mentioned. (3) Discussions are necessarily brief. Here, too, we assume that the literature citations will serve as starting points for the reader to gather much more information on each topic. A later section lists a few key sources of information commonly used by the growing HSI community of practice.

Computation *Before* Full-Scene Data Analysis

Atmospheric Compensation (AC)

AC is the process of converting calibrated at-aperture radiance data to reflectance, $\rho(\lambda)$, for the VNIR/SWIR and to ground-leaving radiance data (GLR) for the LWIR. LWIR GLR data are then converted to emissivity, $\varepsilon(\lambda)$, by temperature/emissivity separation (TES).[5] Though AC is considered primarily the process for getting $\rho(\lambda)$ and $\varepsilon(\lambda)$, it may also be considered an inversion to obtain the atmospheric state captured in the HSI data. Much has been written about AC for HSI (and MSI). Additionally, AC borrows heavily from atmospheric science—another field with an extensive literature.

AC is accomplished via one of two general approaches. (1) In scene methods such as QUAC (Bernstein et al. 2005) or ELM, both for the VNIR/SWIR; or ISAC (Young et al. 2002) for the LWIR. (2) RT models such as MODTRAN.[6] In practice, the RT models are used in conjunction with in-scene data such as atmospheric water vapor absorption band-depth to guide LUT search for estimating transmissivity. Tools such as FLAASH (Adler-Golden et al. 2008) are GUI-driven and combine the use of MODTRAN and the interaction with the data to generate reflectance. The process is similar for the LWIR; AAC is an example of this (Gu et al. 2000).

---

[5]Reflectivity and emissivity are related by Kirchhoff's law: $\varepsilon(\lambda) = 1 - \rho(\lambda)$.

[6]MODTRAN (v5) is extremely versatile and may be used for HSI data from the VNIR through the LWIR.

It is also possible to build a single RT model based tool to ingest LWIR at-aperture radiance data and generate emissivity that essentially eliminates (actually subsumes) the separate TES process.

In-scene AC methods span the range of computational burden/overhead from low (ELM) to moderate/high (QUAC). RT methods, however, can span the gamut from 'simple' LUT generation to increasing the complexity of the RT expressions and numerical analytical techniques used in the model. This is then followed by increasing the complexity of the various interpolation and optimization schemes utilized with the actual remotely sensed data to retrieve reflectance or emissivity. Here, too, when trying to match a physical measurement to modeled data, the entire pantheon of existing and emerging multivariate analysis, optimization, etc., techniques may be utilized.

In a nutshell, quite a bit of AC for HSI is RT-model driven combined with in-scene information. It should also be noted that typical HSI analysis generates one AC solution for each scene. Depending on the spatial dimensions of the scene, its expected statistical variance, or scene-content complexity, one or several solutions may be appropriate. As such, opportunities to expend computational resources utilizing a broad range of algorithmic complexity are many.

Regression Remote Sensing

Regression remote sensing was described above and is only briefly recapped here. It is exemplified by the use of stepwise least-squares regression analysis to select the best bands or combination of bands that correlate one (or more) desired parameters from the data. An example would be foliar chemistry derived by field sampling (followed by laboratory analysis) at the same time an HSI sensor was collecting. Coefficients are generally derived using the actual remotely sensed HSI data (and laboratory analyses) but may be derived using ground-truth point spectrometer data (with sampling characteristics comparable to the airborne HSI sensor; see ASD, Inc. 2012). Computation is required for the statistical analyses that generate coefficients for the model (regression) equation (e.g., an nth-degree polynomial) which will then be applied to the remotely sensed HSI data set. The need for computational resources can vary widely. Developers may draw on a large and growing inventory of techniques for multivariate analysis, optimization, etc., techniques for fitting and for feature selection. The ultimate application of the model to the actual HSI data is generally not algorithmically demanding or computationally complex.

Computation *During* Full-Scene Data Analysis

Data Exploration: PCA, MNF, and ICA

Principal components analysis (PCA), minimum noise fraction (MNF; Green et al. 1988), and independent components analysis (ICA; e.g., Comon 1994) are statistical transformations applied to multivariate data sets such as HSI. They are used to: (1) assess data quality and the presence of measurement artifacts; (2) estimate data dimensionality; (3) reduce data dimensionality (see, e.g., the ENVI® Hourglass; ITT Exelis-VIS 2012); (4) separate/highlight unique signatures within the data; and (5) inspect the data in a space different than its native wavelength-basis representation. Interesting color composite images may be built with PCA and MNF results that draw an analyst's attention to features that would otherwise have been overlooked in the original, untransformed space. Second and higher-order statistics are estimated from the data; an eigendecomposition is applied to the covariance (or correlation) matrix. There is perhaps little frontier left in applying PCA and MNF to HSI. The algorithmic complexity and computational burden of these frequently applied processes is quite low when the appropriate computational method is chosen, such as SVD. A PCA or MNF for a moderately sized HSI data cube completes in under a minute on a typical desktop CPU. ICA is different; it is still an active area of research. Computational burden is very high; on an average workstation, an ICA for a moderately sized HSI data cube could take several hours to complete—depending on the details of the specific implementation of ICA being applied—and data volume.

The second-order statistics (e.g., the covariance matrix and its eigenvectors and eigenvalues) generated by a PCA or an MNF may be used by directed material search algorithms (see below). Thus, these transformations may be applied early on for data inspection/assessment *and* to generate information that will be used later in the analysis flow.

HSI Scene Segmentation/Classification

HSI Is to MSI as Spectral Mixture Analysis (SMA) Is to 'Traditional' MSI Classification

Based on traditional analysis of MSI, it has become customary to classify spectral image data—all types. Traditional scene classification as described in, e.g., Richards and Jia (1999) and Lillesand et al. (2008), is indeed possible with HSI but with caveat. (1) Some of the traditional supervised and unsupervised MSI classification algorithms are unable to take full advantage of the increased information content inherent in the very high dimensional, signature-rich HSI data. They report diminishing returns in terms of classification accuracy after some number of features (bands) is exceeded—absorbing computation time but providing no additional benefit.[7] (2) For HSI, it is better to use tools based on spectral mixture analysis

---

[7]This phenomenon has indeed been demonstrated. It is most unfortunate, however, that it has been used to impugn HSI technology when it is really an issue with poor algorithm selection and a lack of understanding of algorithm performance and of the information content inherent in a spectrum.

(SMA; see, e.g., Adams et al. 1993).[8] SMA attempts to unravel and identify spectral signature information from two or more materials captured in the ground resolution cell that yields a pixel in an HSI data cube. The key to successful application of SMA and/or an SMA-variant is the selection of endmembers. And indeed, this aspect of the problem is the one that has received, in our opinion, the deepest, most creative, and most interesting thinking over the last two decades. Techniques include (but certainly not limited to) PPI (Boardman et al. 1995), N-FINDR (Winter 1999), SMACC (Gruninger et al. 2004; ITT Exelis-VIS 2011), MESMA/Viper Tools (Roberts et al. 1998), and AutoMCU (Asner and Lobell 2000). The need for computational resources varies widely based on the endmember selection method. (3) If you insist on utilizing heritage MSI methods (for which the need for computation also varies according to method utilized), we suggest that you do so to the full range HSI data set, and then repeat with successively smaller spectral subsets and compare results. Indeed, consider simulating numerous MSI sensor data sets with HSI by resampling the HSI down to MSI using the MSI systems' bandpass/spectral response functions. More directly, simulate an MSI data set using best band selection (e.g., Keshava 2004) based on the signature(s) of the class(es) to be mapped. Some best band selection approaches have tended to be computationally intensive, though not all. Best band selection is a continuing opportunity for the role of computation in spectral image analysis.

Additional opportunities for computation include combining spectral- and object-based scene segmentation/classification by exploiting the high spatial resolution content of ground-based HSI sensors.

Directed Material Search

The distinction between HSI and MSI is starkest when considering directed material searching. The higher spectral resolution of HSI, the generation of a spectral signature, the resolution of spectral features, facilitates directed searching for specific materials that may only occur in a few or even one pixel (or even be subpixel in abundance within those pixels). HSI is best suited for searching for—and mapping of—specific materials and this activity is perhaps the most common use of HSI. There is a relationship with traditional MSI scene classification, but there are very important distinctions and a point of departure from MSI to HSI. Traditional classification is indeed material mapping but a family of more capable algorithms can take more advantage of the much higher information content inherent in an HSI spectrum.[9] The following sections describe the various algorithms.

---

[8] Also known as spectral unmixing/linear spectral unmixing (LSU), subpixel analysis, subpixel abundance estimation, etc. The mixed pixel, and the challenges it presents, is a fundamental concept underlying much of the design of HSI algorithms and tools.

[9] These algorithms may also be (and have been) applied to MSI. At some level of abstraction, the multivariate statistical signal processing-based algorithms that form the core HSI processing may be applied to *any* multivariate data set (e.g., MSI, HSI, U.S. Dept. of Labor statistics/demographic data) of *any* dimension greater than 1.

Whole Pixel Matching: Spectral Angle and Euclidean Distance

Whole or single pixel matching is the comparison of two spectra. It is a fundamental HSI function; it is fundamental to material identification: the process of matching a remotely sensed spectrum with a spectrum of a known material (generally computer-assisted but also by visual recognition). The two most common methods to accomplish this are spectral angle (θ) mapping (SAM) and minimum Euclidean distance (MED).[10] Note from the numerator of Eq. 1 that the core of SAM is a dot (or inner) product between two spectra, s_1 and s_2 (the denominator is a product of vector magnitudes); MED is the Pythagorean theorem in n-dimensional space. There are many other metrics; many other ways to quantify distance or proximity between two points in n-dimensional space, but SAM and MED are the most common and their mathematical structure underpins the more sophisticated and capable statistical signal processing based algorithms.

$$\theta = \cos^{-1}\left(\frac{s_1{}^{T} s_2}{\|s_1\| \, \|s_2\|} \right) \qquad (1)$$

Whole pixel, in the present context, refers to the process of matching two spectral signatures; a relatively unsophisticated, simple (but powerful) operation. Ancillary information, such as global second-order statistics or some other estimate of background clutter is not utilized (but is in other techniques; see below). Thus, subpixel occurrences of the material being sought may be missed.

There is little algorithmic or computational complexity required for these fundamental operations—even if combined with statistical testing (e.g., the *t*-test in CCSM of van der Meer and Bakker 1997).

Often, a collection of pixels (spectra) from an HSI data set is assumed to represent the same material (e.g., the soil of an exposed extent of ground). These spectra will not be identical to each other; there will be a range of reflectance values within each band; this variation is physically/chemically real and not due to measurement error. Similarly, rarely is there a single 'library' or 'truth' spectral signature for a given material (gases within normal earth surface temperature and pressure ranges being the notable exception). Compositional and textural variability and complexity dictate that a suite of spectra best characterizes any given substance. This is also the underlying concept to selecting training areas in MSI for scene segmentation with, e.g., maximum likelihood classification (MLC). Thus, when calculating distance, it is sometimes best to use metrics that incorporate statistics (as MLC does). The statistics attempt to capture the shape of the cloud of points in hyperspace and use this in estimating distance—usually between two such clouds. Two examples are the Jeffries-Matusita (JM) distance and transformed divergence

---

[10]Sometimes also referred to as simply 'minimum distance' (MD).

(TD). The reader is referred to Richards and Jia (1999) and Landgrebe (2003) for more on the JM and TD metrics and other distance metrics incorporating statistics. Generally speaking, such metrics require the generation and inversion of covariance matrices. The use of such distance metrics is relatively rare in HSI analysis; they are more commonly applied in MSI analysis.

Statistical Signal Processing: MF and ACE

Two pillars of HSI analysis are the spectral matched filter (MF; Eq. 2)[11] and the adaptive coherence/cosine estimator (ACE; Eq. 3) algorithms (see, e.g., Stocker et al. 1990; Manolakis et al. 2003; Manolakis 2005). In Eqs. 2 and 3, μ is the global mean spectrum, t is the desired/sought target spectrum, x is a pixel from the HSI data, and Γ is the covariance matrix (and thus Γ^{-1} is the matrix inverse). MF and ACE are statistical signal processing based methods that use the data's second order statistics (i.e., covariance or correlation matrices) calculated either globally or adaptively. In some sense, they are a culmination of the basic spectral image analysis concepts and methods discussed up to this point. They incorporate the Mahalanobis distance (which is related to the Euclidean distance) and spectral angle, and they effectively deal with mixed pixels. They are easily described (and derived) mathematically and are analytically and computationally tractable. They operate quickly and require minimal analyst interaction. They execute best what HSI does best: directed material search. Perhaps their only downside is that they work best when the target material of interest does not constitute a significant fraction of the scene thus skewing the data statistics upon which they are based (a phenomenon sometimes called 'target leakage'). But even here, at least for the MF, some work-arounds such as reduced rank inversion of the covariance matrix can alleviate this effect (e.g., Resmini et al. 1997). Excellent discussions are provided in Manolakis et al. (2003), Chang (2003), and Schott (2007).[12]

$$D_{AMF}(x) = D_{MF}(x) = \frac{(t - \mu)^T \Gamma^{-1} (x - \mu)}{(t - \mu)^T \Gamma^{-1} (t - \mu)} \tag{2}$$

$$D_{ACE}(x) = \frac{(t - \mu)^T \Gamma^{-1} (x - \mu)}{\sqrt{(t - \mu)^T \Gamma^{-1} (t - \mu)} \sqrt{(x - \mu)^T \Gamma^{-1} (x - \mu)}} \tag{3}$$

---

[11] There are various names for this algorithm. Some are reinventions of the same technique; others represent methods that are variations on the basic mathematical structure as described in, e.g., Manolakis et al. (2003).

[12] As well as an historical perspective provided by the references cited in these works.

Spectral Signature Parameterization (Wavelets, Derivative Spectroscopy, SSA, ln(ρ))

HSI algorithms (e.g., SAM, MED, MF, ACE, SMA) may be applied, as appropriate, to radiance, reflectance, GLR, emissivity, etc., data. They may also be applied to data that have been pre-processed to, ideally, enhance desirable information while simultaneously suppressing components that do not contribute to spectral signature separation. The more common pre-processing techniques are wavelets analysis and derivative spectroscopy. Other techniques include single scattering albedo (SSA) transformation (Mustard and Pieters 1987; Resmini 1997), continuum removal, and a natural logarithm transformation of reflectance (Clark and Roush 1984).

Other pre-processing includes quantifying spectral shape such as band depth, width, and asymmetry to incorporate in subsequent matching algorithms and/or in an expert system; see, e.g., Kruse (2008).

Implementing the Regression Remote Sensing Equations

As mentioned above, applying the model equation, usually an nth-degree polynomial, to the HSI data is not computationally complex or algorithmically demanding. The computational resources and opportunities are invested in the generation of the regression coefficients.

Single Pixel/Superpixel Analysis

Often, pixels which break threshold following an application of ACE or MF are subjected to an additional processing step. This is often (and rightly) considered the actual material identification process but is largely driven by the desire to identify and eliminate false alarms generated by ACE and MF (and every other algorithm). Individual pixels or the average of several pixels (i.e., superpixels) which pass threshold are subjected to matching against a spectral library and, generally, quite a large library. This is most rigorously performed with generalized least squares (GLS) thus incorporating the scene second-order statistics. This processing step becomes very computationally intensive based on spectral library size and the selection of the number of spectral library signatures that may be incorporated into the solution. It is, nonetheless, a key process in the HSI analysis and exploitation flow.

Anomaly Detection (AD)

We have not said anything to this point about anomaly detection (AD). The definition of anomaly is context-dependent. E.g., a car in a forest clearing is an anomaly; the same car in an urban scene is most likely not anomalous. Nonetheless, the algorithms for AD are similar to those for directed material search; many are based on the second-order statistics (i.e., covariance matrix) calculated from the data. For example, the Mahalanobis distance, an expression with the same mathematical form as the numerator of the matched filter, is an AD algorithm. Indeed, an application of the MF (or ACE) may be viewed as AD particularly if another algorithm will be applied to the pixels that pass a user-defined threshold. The MF, in particular, is known to be sensitive to signatures that are 'anomalous' in addition to the signature of the material actually sought. Stated another way, the MF has a reasonably good probability of detection but a relatively high false alarm rate (depending, of course, on threshold applied to the result). This behavior motivated the development of MTMF (Boardman 1998) as well as efforts to combine the output of several algorithms such as MF and ACE. An image of residuals derived from a spectral mixture analysis will also yield anomalies.

Given the similarity of AD methods to techniques already discussed, we will say no more on this subject. The interested reader is referred to Manolakis et al. (2009), and references cited therein, for more information.

Error Analysis

Error propagation through the entire HSI image chain or even through an application of ACE or MF is still an area requiring additional investigation. Though target detection theory (e.g., Neyman-Pearson [NP] theory; see Tu et al. 1997) may be applied to algorithms that utilize statistics, there is a subtle distinction[13] between algorithm performance based on target-signal to background-clutter ratio (SCR; and modifying this by using different spatial and spectral subsets with which data statistics are calculated or using other means to manipulate the data covariance matrix) and the impact of sensor noise on the fundamental ability to make a radiometric measurement; i.e., the NESR, and any additional error terms introduced by, e.g., AC (yielding the NE$\Delta\rho$). NESR impacts minimum detectable quantity (MDQ) of a material, an HSI system (hardware + algorithms) FOM. An interesting assessment of the impact of signature variability on subpixel abundance estimation is given in Sabol et al. (1992) and Adams and Gillespie (2006). See also Kerekes (2008), Brown and Davis (2006), and Fawcett (2006) for detailed discussions on receiver operating characteristic (ROC) curves[14]—another mechanism used to assess HSI system performance and which also have dependencies on signature variability/target SCR and FOMs such as NESR and NE$\Delta\rho$.

---

[13] And a relationship; i.e., signature variability will have two components contributing to the two probability distribution functions in NP theory: an inherent, real variability of the spectral signatures of materials and the noise in the measurement of those signatures imparted by the sensor.

[14] And area under the ROC curve or AUC.

Computational Scope

Much has been said thus far about the algorithms used in HSI analysis. It is worth pausing to discuss the computational implications of hyperspectral data exploitation and the implementation of the algorithms. A typical hyperspectral imagery cube may be 1,000 lines by 500 samples by 250 bands. That is 500,000 pixels or spectra. And though most HSI sensors are 12 or 14 bit systems, the data are handled as 16-bit information. Thus this example data cube is 2×10^9 bits or 250 megabytes. Today, this data cube is small in comparison to the random access memory (RAM) available in modern computers. If read sequentially from RAM to the CPU this operation may take less than 0.04 s. But this is a naïve assessment as the number of operations that must take place, the order in which the data must be read, programming language applied, and the latencies between storage, memory, cache, and CPU must be considered. Let's take a quick look at the order of operations required for a simple hyperspectral algorithm.

Using the above data cube size as an example, a simple calculation of Euclidean distance requires a subtraction of one pixel vector from a reference vector (250 operations), a square of the elements of the result (250 operations), a sum of the vector (249 operations), and a square root of the total (1 operation). This gives 750 operations for each pixel leading to 375 million operations to calculate Euclidean distance for one reference spectrum. This is on the order of n operations where $n = \#$ of pixels x $\#$ of bands. This can quickly escalate as the order of operations for more complex algorithms can approach n^2 operations (endmember decomposition) or even n^3 operations (non-linear methods).

In an ideal world with CPUs reporting performance in the 100 GFLOP range, calculation time would appear to be trivial. But simply adding a 1 microsecond delay to any of these operations results in seconds of latency. In assessing expected performance of these algorithms it is insufficient to compare simple CPU or even GPU reported processing performance. Other latencies of the system, memory access and bandwidth, cache misses, memory and storage read and write speeds, all contribute to the problem and must be assessed.

Interim Summary

Successful HSI analysis is based on the application of specialized algorithms deeply informed by a detailed understanding of the physical, chemical, and radiative transfer processes of the scenario for which the imaging spectroscopy data are acquired. HSI data are significantly more than a seemingly indecipherable collection of points in a high dimensional hyperspace to which an endless mish mash of methods from electrical engineering, signal processing, multivariate analysis, and optimization theory may be blindly applied as a substitute for any and all understanding of the underlying nature and structure of the data and of the objects for which the data were acquired. Apply a technique if its underlying assumptions are met by the HSI data and/or the nature and structure of the HSI data *and* the underpinning physical, chemical, and radiative transfer processes are amenable to the information extraction capabilities of the method.

Miscellaneous Topics

There are many other topics that could be discussed; some commonly applied, others still under development or not yet widely utilized. Topics in the former category include: dimensionality reduction and/or data volume reduction (beyond PCA and MNF); product generation via fusion with lidar and SAR, pan-sharpening, georeferencing, and orthorectification; scene/data modeling and simulation with, e.g., DIRSIG (Schott 2007) and FASSP (Kerekes 2012), and spectral signature modeling. Topics in the later category include: topological methods (Basener et al. 2007); expert systems (Clark et al. 2003; Kruse et al. 1993; Kruse and Lefkoff 1993), genetic algorithms (Harvey et al. 2002), support vector machines (SVMs), Bayesian model averaging (BMA; e.g., Burr and Hengartner 2006); spatial *and* spectral data analysis (Resmini 2012); parallel processing/multicore processing/high performance computing; computer-assisted/analyst interactive data analysis and exploration, and visual analytics; and scientific databases ("big data") and data mining. The interested reader may readily find information on these and many other topics in the scientific literature.

3 A Note to Developers and What's Next

A new technique should be unique, stable, and robust. Its performance should not be easily bested by a skilled, experienced analyst applying the well known, well established toolbox of existing techniques to, say, different spatial and spectral subsets of the HSI data set or after utilizing some simple pre-processing methods (see, e.g., Funk et al. 2001) and/or by simply using the existing workhorse algorithms and tools in sequence and combining the results. Developers are thus urged to: (1) rigorously and honestly compare the performance of their new method with the existing suite of standard tools in the field; (2) apply their new method to a wide diversity of real remotely sensed data and not simply tune algorithm performance for the data set used for development and testing; i.e., honestly probe the technique's performance bounds; and (3) perhaps most importantly, carefully review (and cite) the literature to avoid reinventing the wheel. And it cannot be stressed too strongly: computation is not a substitute for a deeper understanding of the nature of HSI data and practical knowledge of the problem and its setting for which the data have been acquired. That being said and emphasized, the next several sections discuss how computational resources can be applied.

3.1   Desktop Prototyping and Processing Peril

Many HSI practitioners develop new methods and algorithms out of necessity. Solving unique problems requires development or modification of algorithms for specific needs. The availability of desktop programming and mathematical tools such as Matlab or IDL has increased our productivity tremendously. These commercially available tools abstract complex algorithms into simple function calls for easy implementation. This is not without peril. Although the majority of new algorithm development applies sound fundamentals in regards to phenomenology, there is a need to understand the computational complexities of these approaches. A quick perusal of the help files of desktop prototyping tools such as Matlab or IDL for a simple function such as matrix inverse, will lead to discussion and examples of non-exact solutions and warnings of singular matrices. Since our fundamental problem is that of remote sensing inversion (Twomey 1977), we must expect that our computational results can yield non-exact or non-physical solutions. In addition, a naïve application of these functions may lead to significant computational issues such as rounding and truncation due to machine precision. A simple computational example ($b = Ax$) for solving a set of linear equations in Matlab is illustrated below.

```
>> A = magic(3)
A =
8 1 6
3 5 7
4 9 2
>> A(:,1) = zeros(1,3)
A =
0 1 6
0 5 7
0 9 2
>> b = [1;2;5];
>> x = A\b
Warning: Matrix is singular to working precision.
x =
Inf
0.584905660377358
-0.132075471698113
>> x = pinv(A)*b
x =
0
0.531943065210195
-0.015557762330354
>> A*x-b
ans =
-0.561403508771931
0.550810989738495
-0.243627937768954
```

In this example, we attempt to solve a set of linear equations using both Matlab's '\' operator (matrix inverse) and 'pinv' (pseudoinverse). Both solutions attempt a least squares solution. In the case of matrix inverse the solution goes to infinity and gives a warning of a singular matrix. Attempting a pseudoinverse leads to a solution, but comparison to the original vector **b** leads to a surprising result. The function did not fail, but the calculation did—and without warning.

The matrix inverse operation is key to many steps of the HSI analysis process and necessitates a check of both data quality and validity of results. This example illustrates sensitivity of a solution to the methods and values applied. As a practical exercise, one may choose to attempt a spectral unmixing method with artifact-laden or poorly calibrated data, e.g., bad/noisy bands, bands of all zeros, etc., and study the stability and physical implications of the unmixing model and its residuals when applying a pseudoinverse method.

Another critical operation to many algorithms is calculation of the covariance matrix. This calculation is a relatively straight forward combination of subtraction operations and array multiplication. While these operations present no inherent computational issues, the choice and quantity of pixels used for covariance estimation are critical. In regards to selection of which pixels to use, an assumption in calculating covariance for the matched filter is that it represents a homogenous background population. Target materials or anomalies present in the covariance estimation significantly degrade performance of the matched filter. In regards to quantity, the size of the background population for covariance estimation can suffer from two pitfalls: (1) the pixels chosen should represent the variance of the background data. Using pixels which are too similar or too varied (i.e., contain target materials or anomalies) will again degrade performance of the algorithm. (2) the quantity of pixels chosen should be sufficient to avoid computational issues of inverting a singular matrix. A good rule of thumb is to estimate the covariance with at least 10-times the number of data dimensions.

Significant effort has been made to ensure the computational accuracy of these methods and their implementation in software packages. Many of the desktop packages utilize the well known BLAS (Basic Linear Algebra Subprograms) and LAPACK (Linear Algebra Package) algorithm libraries first developed in the 1980s and 1990s and continually updated (Anderson 1999). These libraries are highly efficient implementations of numerical linear algebra methods for single and double precision and real and complex calculations. Functions also return flags indicating some measure of validity of the returned result. A basic understanding of these methods and their implementation in desktop computing applications should not be overlooked. This understanding parallels the deeper understanding of HSI data and practical knowledge of the problem as stated previously.

3.2 Automated Processing and Time Critical Applications

Discussions so far have focused on analysis methods suited for manual or analyst-interactive processing of individual HSI data sets. As the number of hyperspectral sensors and thus data increase in both military and civilian applications, the need for automated processing increases. Although much can be said about the complexities of this particular remote sensing problem, the need to automatically process data for anomaly detection or directed material search remains. In general, automated hyperspectral processing is driven by two circumstances: (1) the availability of suitable data analysts; and (2) the need for time critical analysis. In regards to 1, we believe it is safe to propose that the growth of HSI data will always exceed the availability of suitable analysts. Given that, automated processing for a portion, if not for all of HSI analysis, is necessary to support the limited availability of HSI analysts.

In regards to 2, hyperspectral sensors as a reconnaissance and surveillance tool seek to provide information and not just data to the appropriate first responders and decision makers. It would be naïve of us to consider only scientists and engineers as the sole consumers of such information. Because of this, automated processing to discover specific types of information is a necessity. As experts in the methodologies of HSI analysis, it is up to us to develop suitable methods of automated processing for the non-expert user and to thoroughly understand and explain the constraints in which that automated processing is valid. Automated processing is there to support the time critical nature of a specific mission or objective—e.g. military operations or disaster support. Choosing appropriate algorithms, specific target libraries, and providing some method of data/processing quality assurance and confidence is absolutely necessary.

Time critical analysis is driven by a need for information as soon as possible after the data are collected. This can be either in-flight or post-flight. An in-flight scenario requires on-board processing in which there may or may not be analyst on-board; e.g. UAV. In a post-flight scenario, multiple analysts and a mission specific set of computing hardware and software may be available. In both cases algorithm and target library selection remain critical. Typically, automated processes are studied in detail for specific target libraries before implementation in an actual data collection operation. An example process includes the following steps:

1. Data pre-processing: This step generally brings the data from DN to a calibrated radiance. A check of the data for data quality prior to processing is performed. Bad bands and pixels may be removed. Geo-registration may be performed.
2. Atmospheric correction: This step converts at-sensor radiance to reflectance. This may be an in-scene or RT/modeled method.
3. Target Detection: In this step a statistical detector such as MF or ACE is applied using the target spectral library. Individual target detection planes are created.
4. Thresholding: Using the detection planes and a predetermined threshold, pixels above the threshold are selected as possible target materials. Spatial operations are performed to generate discrete regions of interest (ROI).
5. Identification: To confirm results of detection, individual spectra from the ROIs are compared to a larger set of target materials. Comparisons are made using various methods such as SAM, MED, or step-wise linear regression. This is the spectroscopy step of HSI analysis. Score values are then generated for each ROI.

An individual on a desktop computer may take several minutes to analyze a single data set following these steps. Implemented as a fully automated process on a GPU and using a target library of a hundred materials, this can be completed is several seconds (Brown et al. 2012).

This brings us to another scenario which drives time critical analysis and that is real time or near real time processing. There appears to be some misconception that on-board analysis necessitates real time processing of HSI data. The mention of real time processing usually leads to discussion of what real time processing is. In the present context, we use real time and near real time processing interchangeably. We define near real time processing to be automated processing with very low processing latency; e.g., a few seconds. In other words, once an HSI data cube is collected, it is then processed in an amount of time less than or equal to the collection time. Practical experience shows us that the difference between a few seconds or even a few minutes of processing latency is insignificant in most applications where the HSI sensor is the primary or only data collector. The fact that the sensor platform observed a location one or more seconds ago has little bearing on the ability of the sensor or processing algorithm's ability to confidently perform a directed material search. The critical requirement is that on-board processing keeps pace with the data collection rate of the HSI sensor such that the initial processing latency allows the sensor system (i.e., hardware plus processing) to provide relevant information while it remains in its desired operating area.

A more stringent real time processing requirement occurs when HSI data are combined for data or information fusion with sensors that collect and process data that have temporal relevance such as motion imagery. In this case, the materials of interest may have a persistent signature, but the activities identified in the motion imagery are fleeting. It is now critical to overlay track or cue information onto broadband imagery such that an analyst/operator can associate spectral information with motion based activity. Processing latencies of more than just a few seconds would be unacceptable for real time vehicle tracking that combines spectral and motion imagery.

Real time HSI processing systems and algorithms have been pursued over the past several years with varying success (Stevenson et al. 2005; Chang 2013; Brown et al. 2012). The availability of inexpensive high performance computing hardware (GPU, DSP, FPGA) and their associated development environments facilitate the migration of HSI algorithms to embedded computing applications. In recent years the migration of HSI algorithms to GPUs has been researched and widely published on (e.g., Morgenstern and Zell 2011; Trigueros-Espinosa et al. 2011; Winter and Winter 2011). More complex algorithms to include non-linear methods and HSI georeferencing have also found significant performance improvement on GPUs (Campana-Olivo and Manian 2011; Opsahl et al. 2011). It is simply a matter of time before our most reliable and robust HSI algorithms are operating as ubiquitous automated processors.

3.3   A New Paradigm: Big Data

Up to this point we have largely considered analyst-interactive analysis of individual HSI data sets. This is either a desktop process conducted by an analyst, or possibly a near real time system processing data cubes as they are collected. A new paradigm in data analysis exists that must now be considered for spectral processing and exploitation. To motivate the reader we pose the following questions:

1. Consider the scope of your spectral data holdings. If you had the ability to process and analyze groupings of data or the entire collection/campaign of data in minutes, would you want to?
2. Have you ever considered the temporal or spatial evolution of material signatures, atmospheric effects, data covariance, or any other aspects of your hyperspectral information across years of collected data?
3. Can you now analyze more than one data cube simultaneously and jointly?
4. If Google had access to your data, how would they store, process, analyze, distribute, and study it?

Most of us are familiar with Google and maybe somewhat familiar with cloud computing. What most of us are not familiar with are the concepts of Big Data and the volume of information it represents. Years ago, when we considered the difficulty in processing large hyperspectral data sets, our concepts of big data were limited by our processing ability on a single CPU or possibly across multiple CPUs in a homogenous compute cluster. Today, Big Data represents the vast amount of structured and unstructured digital data that simply exist on computers and servers the world over. Big Data is of such concern to the commercial, business, and defense communities, in March 2012 the Office of the President of the United States announced the "Big Data Research and Development Initiative[15]". This initiative funds efforts across the U.S. Government to research and develop techniques and methodologies to process and exploit extremely large data holdings. This includes intelligence, reconnaissance, and surveillance data from DoD, the vast holdings of earth observation and remote sensing data from NASA, and large data holdings across NIH, DOE and many other government agencies.

The first step in approaching the Big Data problem is an understanding of existing tools and methodologies for a distributed computing environment. This begins with Mapreduce developed by Google. Mapreduce is a programming model and implementation for processing large data sets (Dean and Ghemawat 2008). Programs written in the Mapreduce construct are automatically parallelized and can be reliably executed on large distributed heterogeneous systems. Using the Mapreduce model allows simplified development of parallel processing methods across thousands of distributed computers. Mapreduce is the basis of the production

---

[15]http://www.whitehouse.gov/blog/2012/03/29/big-data-big-deal

indexing system supporting the Google web search (Dean and Ghemawat 2008) and has been found effective in various applications such as machine learning, bioinformatics, astrophysics, and cyber-security (Lin et al. 2010).

Mapreduce has been implemented in the open-source application Hadoop, developed by the Apache Software Foundation.[16] Hadoop has become the preferred solution for Big Data analytics and is in use by Google, Yahoo, IBM, Facebook, and others (Burlingame 2012). Hadoop implements distributed computing and distributed file system elements with a Java programming interface to allow for the development of distributed computing environments. A Hadoop implementation is available to users of Amazon Web Services as Amazon Elastic MapReduce (EMR).[17] EMR provides access to a user configurable number of compute nodes and charges a fee based on compute capacity needed. Amazon has effectively and inexpensively provided supercomputer access to any individual, company, or government.

Mapreduce has created a new kind of supercomputer for Big Data analysis (McMillan 2012). In this context, HSI analysis must be viewed no longer in terms of full-scene analysis, but full-campaign analysis, or full-regional analysis, or fully integrated temporal-spatial analysis. It is now up to us to integrate our practical knowledge of HSI analysis with the computational resources available to anyone with access to a computer and the internet.

3.4   Where to Find More Information: The HSI Community of Practice

HSI remote sensing is an established, active field of research and practical application with a large and growing body of literature. Practitioners and would-be contributors have many resources at their disposal for research on previous work and for communication of results. Scientific journals include Remote Sensing of Environment, the International Journal of Remote Sensing, the IEEE[18] Transactions on Geoscience and Remote Sensing, and the IEEE Geoscience and Remote Sensing Letters. Scientific associations include the Society of Photo-optical Instrumentation Engineers (SPIE), IEEE, the American Society of Photogrammetry and Remote Sensing (ASPRS), and the American Geophysical Union (AGU). Each society has a host of journals, both peer reviewed and non-reviewed, and major symposia at which results are communicated. HSI remote sensing is a vigorous community of practice and one in which government, private sector, and academic institutions participate. A wealth of information about HSI is also available on the World Wide Web.

---

[16]http://hadoop.apache.org/, last accessed May 8, 2012.

[17]http://aws.amazon.com/elasticmapreduce/

[18]Institute of Electrical and Electronics Engineers.

A.1   Appendix: Acronyms, Symbols, and Abbreviations Table

θ	Spectral angle
(superscript) T	Transpose
AAC	Autonomous atmospheric compensation
AC	Atmospheric compensation
ACE	Adaptive coherence/cosine estimator
AD	Anomaly detection
AGU	American Geophysical Union
ASD	Analytical Spectral Devices (formerly)
ASPRS	American Society of Photogrammetry and Remote Sensing
AUC	Area under the (ROC) curve
AutoMCU	Automated Monte Carlo unmixing
BLAS	Basic Linear Algebra Subprograms
BMA	Bayesian modeling averaging
CCSM	Cross correlogram spectral matching
$\cos$, $\cos^{-1}$	Cosine, inverse cosine (arccosine)
CPU	Central processing unit
DoD	U.S. Department of Defense
DOE	U.S. Department of Energy
DIRSIG	Digital imaging and remote sensing image generation model
DN	Digital number
DSP	Digital signal processor
ELM	Empirical line method
EMR	Elastic MapReduce
ENVI	Environment for Visualizing Images
FASSP	Forecasting and analysis of spectroradiometric system performance model
FLAASH	Fast line-of-sight atmospheric adjustment of spectral hypercubes
FOM	Figure of merit
FPGA	Field-programmable gate array
GFLOP	Giga-floating point operations
GLR	Ground-leaving radiance
GPU	Graphics processing unit
GUI	Graphical user interface
HSI	Hyperspectral imagery
ICA	Independent components analysis
IDL	Interactive Data Language
IEEE	Institute of Electrical and Electronics Engineers
ISAC	In-scene atmospheric compensation
JM	Jeffries-Matusita
LAPACK	Linear Algebra Package
ln()	Natural logarithm
LSU	Linear spectral unmixing
LUT	Lookup table
LWIR	Longwave infrared
MD	Minimum distance

(continued)

(continued)

MDQ	Minimum detectable quantity
MED	Minimum Euclidean distance
MESMA	Multiple endmember spectral mixture analysis
MF	Matched filter
MLC	Maximum likelihood classification
MNF	Minimum noise fraction transform
MODTRAN	Moderate resolution transmission tool
MSI	Multispectral imagery
MTMF	Mixture tuned matched filtering
MWIR	Midwave infrared
NaN	Not a number
NASA	U.S. National Aeronautics and Space Administration
NESR	Noise equivalent spectral radiance
NEΔT	Noise equivalent change in temperature
NE$\Delta\varepsilon$	Noise equivalent change in emissivity
NE$\Delta\rho$	Noise equivalent change in reflectance
N-FINDR	N-finder; spectral endmember finder tool
NIH	U.S. National Institutes of Health
NP	Neyman-Pearson
PC	Principal components (shortened notation for PCA)
PCA	Principal components analysis
PCR	Principal components regression
PLSR	Partial least squares regression
QUAC	Quick atmospheric correction
RAM	Random access memory
ROC	Receiver operating characteristic curve
ROI	Region of interest
RT	Radiative transfer
SAM	Spectral angle mapper
SAR	Synthetic aperture radar
SCR	Signal to clutter ratio
SEBASS	Spatially enhanced broadband array spectrograph system
SMA	Spectral mixture analysis
SMACC	Sequential maximum angle convex cone
SME	Subject matter expert
SPIE	Society of Photo-optical Instrumentation Engineers
SSA	Single scattering albedo
SVM	Support vector machines
SWIR	Shortwave infrared
T	Temperature
t	Target spectrum (see Eqs. 2 and 3)
TD	Transformed divergence

(continued)

(continued)

TES	Temperature/emissivity separation
TIR	Thermal infrared
UAV	Unmanned aerial vehicle
VNIR	Visible/near-infrared
x	Scene spectrum (see Eqs. 2 and 3)
Γ, Γ^{-1}	Covariance matrix, inverse of the covariance matrix
$\varepsilon(\lambda)$	Emissivity
λ	Wavelength
μ	Mean spectrum (see Eqs. 2 and 3)
μm	Micrometer
$\rho(\lambda)$	Reflectance

References

Adams JB, Gillespie AR (2006) Remote sensing of landscapes with spectral images: a physical modeling approach. Cambridge University Press, Cambridge, 362 p

Adams JB, Smith MO, Gillespie AR (1993) Imaging spectroscopy: interpretation based on spectral mixture analysis. In: Pieters CM, Englert PAJ (eds) Remote geochemical analysis: elemental and mineralogical composition, vol 4, Topics in remote sensing. Cambridge University Press, Cambridge, pp 145–166

Adler-Golden S, Berk A, Bernstein LS, Richtsmeier S, Acharya PK, Matthew MW, Aderson GP, Allred CL, Jeong LS, Chetwynd JH (2008) FLAASH, a MODTRAN4 atmospheric correction package for hyperspectral data retrieval and simulations. ftp://popo.jpl.nasa.gov/pub/docs/workshops/98_docs/2.pdf. Last accessed 29 Jan 2012

Anderson E, Bai Z, Bischof C, Blackford S, Demmel J, Dongarra J, Du Croz J, Greenbaum A, Hammarling S, McKenney A, Sorensen D (1999) LAPACK users' guide, 3rd edn. Society for Industrial and Applied Mathematics, Philadelphia

Asner GP, Lobell DB (2000) A biogeophysical approach for automated SWIR unmixing of soils and vegetation. Remote Sens Environ 74:99–112

Basener B, Ientilucci EJ, Messinger DW (2007) Anomaly detection using topology. In: Proceedings of SPIE, algorithms and technologies for multispectral, hyperspectral, and ultraspectral imagery XIII, vol 6565. Orlando, April 2007

Bernstein LS, Adler-Golden SM, Sundberg RL, Levine RY, Perkins TC, Berk A, Ratkowski AJ, Felde G, Hoke ML (2005) Validation of the QUick atmospheric correction (QUAC) algorithm for VNIR-SWIR multi- and hyperspectral imagery. In: Shen SS, Lewis PE (eds) Proceedings of the SPIE, algorithms and technologies for multispectral, hyperspectral, and ultraspectral imagery XI, vol 5806. Orlando, 28 Mar–1 Apr 2005, pp 668–678

Boardman JW (1998) Leveraging the high dimensionality of AVIRIS data for improved subpixel target unmixing and rejection of false positives: mixture tuned matched filtering. In: Green RO (ed) Proceedings of the 7th JPL geoscience workshop, NASA Jet Propulsion Laboratory, pp 55–56

Boardman JW, Kruse FA, Green RO (1995) Mapping target signatures via partial unmixing of AVIRIS data. In: Summaries, fifth JPL airborne earth science workshop, NASA Jet Propulsion Laboratory Publication 95–1, vol 1, pp 23–26

Brown CD, Davis HT (2006) Receiver operating characteristics curves and related decision measures: a tutorial. Chemometr Intell Lab Syst 80:24–38. doi:10.1016/j.chemolab.2005.05.004

Brown MS, Glaser E, Grassinger S, Slone A, Salvador M (2012) Proceeding of SPIE 8390. Algorithms and technologies for multispectral, hyperspectral, and ultraspectral imagery XVIII 839018, 8 May 2012. doi:10.1117/12.918667

Burlingame N (2012) The little book of big data. New Street Communications, LLC., Wickford, 590 p

Burr T, Hengartner N (2006) Overview of physical models and statistical approaches for weak gaseous plume detection using passive infrared hyperspectral imagery. Sensors 6: 1721–1750 (http://www.mdpi.org/sensors)

Campana-Olivo R, Manian V (2011) Parallel implementation of nonlinear dimensionality reduction methods applied in object segmentation using CUDA in GPU. In: Proceedings of SPIE 8048, algorithms and technologies for multispectral, hyperspectral, and ultraspectral imagery XVII, 80480R, 20 May 2011, doi:10.1117/12.884767

Campbell JB (2007) Introduction to remote sensing, 4th edn. The Guilford Press, New York, 626 p

Chang C-I (2003) Hyperspectral imaging. Techniques for spectral detection and classification. Kluwer/Plenum Publishers, New York, 370 p

Chang C-I (2013) Real time hyperspectral image processing: algorithm architecture and implementation. Springer (in press) ISBN 978-1-4419-6186-0, 490 p

Clark RN, Roush TL (1984) Reflectance spectroscopy: quantitative analysis techniques for remote sensing applications. J Geophys Res 89(B7):6329–6340

Clark RN, Swayze GA, Live KE, Kokaly RF, Sutley SJ, Dalton JB, McDougal RR, Gent CA (2003) Imaging spectroscopy: earth and planetary remote sensing with the USGS Tetracorder and expert systems. J Geophys Res 108(E12):5131. doi:10.1029/2002JE001847

Comon P (1994) Independent component analysis, a new concept? Signal Process 36:287–314

Dean J, Ghemawat S (2008) Mapreduce: simplified data processing on large clusters. Comm ACM 51(1):107–113

Eismann MT (2012) Hyperspectral remote sensing. SPIE Press, Bellingham, 725 p

ITT Exelis-VIS (2011) ENVI tutorial: using SMACC to extract endmembers. www.exelisvis.com/portals/0/tutorials/envi/SMACC.pdf. Last accessed 12 Feb 2012

ITT Exelis-VIS (2012) http://www.exelisvis.com/language/en-us/productsservices/envi.aspx. Last accessed 17 Oct 2012

Fawcett T (2006) An introduction to ROC analysis. Pattern Recogn Lett 27:861–874. doi:10.1016/j.patrec.2005.10.010

Feilhauer H, Asner GP, Martin RE, Schmidtlein S (2010) Brightness-normalized partial least squares regression for hyperspectral data. J Quant Spectrosc Radiat Transfer 111:1947–1957

FLAASH http://www.spectral.com/remotesense.shtml#FLAASH. Last accessed 29 Jan 2012

Funk CC, Theiler J, Roberts DA, Borel CC (2001) Clustering to improve matched filter detection of weak gas plumes in hyperspectral thermal imagery. IEEE T Geosci Remote Sens 39(7): 1410–1420

Green AA, Berman M, Switzer P, Craig MD (1988) A transformation for ordering multispectral data in terms of image quality with implications for noise removal. IEEE T Geosci Remote Sens 26(1):65–74

Gruninger J, Ratkowski AJ, Hoke ML (2004) The Sequential Maximum Angle Convex Cone (SMACC) endmember model. In: Shen SS, Lewis PE (eds) Proceedings of the SPIE, algorithms for multispectral and hyper-spectral and ultraspectral imagery, vol 5425–1. Orlando, April 2004

Gu D, Gillespie AR, Kahle AB, Palluconi FD (2000) Autonomous atmospheric compensation (AAC) of high-resolution hyperspectral thermal infrared remote-sensing imagery. IEEE T Geosci Remote Sens 38(6):2557–2570

Hackwell JA, Warren DW, Bongiovi RP, Hansel SJ, Hayhurst TL, Mabry DJ, Sivjee MG, Skinner JW (1996) LWIR/MWIR imaging hyperspectral sensor for airborne and ground-based remote sensing. In: Proceedings of the SPIE, vol 2819, pp 102–107

Hapke B (1993) Theory of reflectance and emittance spectroscopy. Cambridge University Press, Cambridge, 455 p

Harvey NR, Theiler J, Brumby SP, Perkins S, Szymanski JJ, Bloch JJ, Porter RB, Galassi M, Young AC (2002) Comparison of GENIE and conventional supervised classifiers for multispectral image feature extraction. IEEE T Geosci Remote Sens 40(2):393–404

Hecht E (1987) Optics, 2nd edn. Addison-Wesley Publishing Company, Reading, 676 p

ASD Inc. (2012) http://www.asdi.com/. Last accessed 29 Jan 2012

Jensen JR (2007) Remote sensing of the environment: an earth resource perspective, 2nd edn. Prentice Hall Series in Geographic Information Science, Upper Saddle River, 608 p

Kerekes JP (2008) Receiver operating characteristic curve confidence intervals and regions. IEEE Geosci Remote Sens Lett 5(2):251–255

Kerekes JP (2012) http://www.cis.rit.edu/people/faculty/kerekes/fassp.html. Last accessed 2 Feb 2012

Keshava N (2004) Distance metrics and band selection in hyperspectral processing with application to material identification and spectral libraries. IEEE T Geosci Remote Sens 42(7):1552–1565

Kokaly RF, Clark RN (1999) Spectroscopic determination of leaf biochemistry using band-depth analysis of absorption features and stepwise multiple linear regression. Remote Sens Environ 67:267–287

Kruse FA (2008) Expert system analysis of hyperspectral data. In: Shen SS, Lewis PE (eds) Proceedings of the SPIE, algorithms and technologies for multispectral, hyperspectral, and ultraspectral imagery XIV, vol 6966, doi:10.1117/12.767554

Kruse FA, Lefkoff AB (1993) Knowledge-based geologic mapping with imaging spectrometers: remote sensing reviews, special issue on NASA Innovative Research Program (IRP) results, vol 8, pp 3–28. http://www.hgimaging.com/FAK_Pubs.htm. Last accessed 29 Jan 2012

Kruse FA, Lefkoff AB, Dietz JB (1993) Expert system-based mineral mapping in northern Death Valley, California/Nevada using the Airborne Visible/Infrared Imaging Spectrometer (AVIRIS): remote sensing of environment, special issue on AVIRIS, May–June 1993, vol 44, pp 309–336. http://www.hgimaging.com/FAK_Pubs.htm. Last accessed 29 Jan 2012

Landgrebe DA (2003) Signal theory methods in multispectral remote sensing. Wiley-Interscience/Wiley, Hoboken, 508 p

Lillesand TM, Kiefer RW, Chipman JW (2008) Remote sensing and image interpretation, 6th edn. Wiley, New York, 756 p

Lin H, Archuleta J, Ma X, Feng W, Zhang Z, Gardner M, (2010) MOON: MapReduce on opportunistic environments. In: Proceedings of the 19th ACM international symposium on high performance distributed computing, ACM, New York

Manolakis D (2005) Taxonomy of detection algorithms for hyperspectral imaging applications. Opt Eng 44(6):1–11

Manolakis D, Marden D, Shaw GA (2003) Hyperspectral image processing for automatic target detection applications. MIT Lincoln Lab J 14(1):79–116

Manolakis D, Lockwood R, Cooley T, Jacobson J (2009) Is there a best hyperspectral detection algorithm? In: Shen SS, Lewis PE (eds) Algorithms and technologies for multispectral, hyperspectral, and ultraspectral imagery XV, vol 7334. Orlando, doi:http://dx.doi.org/10.1117/12.816917, 16 p

McMillan R (2012) Project moon: one small step for a PC, one giant leap for data. http://www.wired.com/wiredenterprise/2012/05/project_moon/. Last accessed 8 May 2012

MODTRAN5 http://www.modtran.org/. Last accessed 29 Jan 2012

Morgenstern J, Zell B (2011) GPGPU-based real-time conditional dilation for adaptive thresholding for target detection. In: Proceedings of SPIE 8048, algorithms and technologies for multispectral, hyperspectral, and ultraspectral imagery XVII, 80480P, 20 May 2011, doi:10.1117/12.890851

Mustard JF, Pieters CM (1987) Abundance and distribution of ultramafic microbreccia in moses rock dike: quantitative application of mapping spectroscopy. J Geophys Res 92(B10): 10376–10390

Opsahl T, Haavardsholm TV, Winjum I (2011) Real-time georeferencing for an airborne hyperspectral imaging system. In: Proceedings of SPIE 8048, algorithms and technologies for multispectral, hyperspectral, and ultraspectral imagery XVII, 80480S, 20 May 2011, doi:10.1117/12.885069

Resmini RG (1997) Enhanced detection of objects in shade using a single-scattering albedo transformation applied to airborne imaging spectrometer data. The international symposium on spectral sensing research, CD-ROM, San Diego, 7 p

Resmini RG (2012) Simultaneous spectral/spatial detection of edges for hyperspectral imagery: the HySPADE algorithm revisited. In: Shen SS, Lewis PE (eds) Proceedings of the SPIE, algorithms and technologies for multispectral, hyperspectral, and ultraspectral imagery XVIII, vol 8390. Baltimore, 23–27 April 2012, doi:http://dx.doi.org/10.1117/12.918751, 12 p

Resmini RG, Kappus ME, Aldrich WS, Harsanyi JC, Anderson ME (1997) Mineral mapping with Hyperspectral Digital Imagery Collection Experiment (HYDICE) sensor data at Cuprite, Nevada, U.S.A. Int J Remote Sens 18(7):1553–1570. doi:10.1080/014311697218278

Richards JA, Jia X (1999) Remote sensing digital image analysis, an introduction, 3rd, revised and enlarged edition. Springer, Berlin, 363 p

Roberts DA, Gardner M, Church R, Ustin S, Scheer G, Green RO (1998) Mapping Chaparral in the Santa Monica Mountains using multiple endmember spectral mixture models. Remote Sens Environ 65:267–279

Sabol DE, Adams JB, Smith MO (1992) Quantitative subpixel spectral detection of targets in multispectral images. J Geophys Res 97:2659–2672

Schaepman-Strub G, Schaepman ME, Painter TH, Dangel S, Martonchik JV (2006) Reflectance quantities in optical remote sensing—definitions and case studies. Remote Sens Environ 103:27–42

Schott JR (2007) Remote sensing: the image chain approach, 2nd edn. Oxford University Press, New York, 666 p

Solé JG, Bausá LE, Jaque D (2005) An introduction to the optical spectroscopy of inorganic solids. Wiley, Hoboken, 283 p

Stevenson B, O'Connor R, Kendall W, Stocker A, Schaff W, Alexa D, Salvador J, Eismann M, Barnard K, Kershenstein J (2005) Design and performance of the civil air patrol ARCHER hyperspectral processing system. In: Proceedings of SPIE, vol 5806, p 731

Stocker AD, Reed IS, Yu X (1990) Multi-dimensional signal processing for electro-optical target detection. In: Signal and data processing of small targets 1990, Proceedings of the SPIE, vol 1305, pp 218–231

Trigueros-Espinosa B, Vélez-Reyes M, Santiago-Santiago NG, Rosario-Torres S (2011) Evaluation of the GPU architecture for the implementation of target detection algorithms for hyperspectral imagery. In: Proceedings of SPIE 8048, algorithms and technologies for multispectral, hyperspectral, and ultraspectral imagery XVII, 80480Q, May 20 2011, doi:10.1117/12.885621

Tu TM, Chen C-H, Chang C-I (1997) A least squares orthogonal subspace projection approach to desired signature extraction and detection. IEEE T Geosci Remote Sens 35(1):127–139

Twomey S (1977) Introduction to the mathematics of inversion and indirect measurements. Development in geomathematics, no. 3. Elsevier Scientific Publishing, Amsterdam, (republished by Dover Publ., 1996), 243 p

van Der Meer F, Bakker W (1997) CCSM: cross correlogram spectral matching. Int J Remote Sens 18(5):1197–1201. doi:10.1080/014311697218674

Winter ME (1999) N-FINDR: an algorithm for fast autonomous spectral end-member determination in hyperspectral data. In: Descour MR, Shen SS (eds) Proceedings of the SPIE, imaging spectrometry V, vol 3753. Denver, 18 July 1999, pp 266–277, doi:10.1117/12.366289

Winter ME, Winter EM (2011) Hyperspectral processing in graphical processing units. In: Proceedings of SPIE 8048, algorithms and technologies for multispectral, hyperspectral, and ultraspectral imagery XVII, 80480O, 20 May 2011, doi:10.1117/12.884668

Young SJ, Johnson RB, Hackwell JA (2002) An in-scene method for atmospheric compensation of thermal hyperspectral data. J Geophys Res 107(D24):4774. doi:10.1029/2001JD001266, 20 p

Toward Understanding Tornado Formation Through Spatiotemporal Data Mining

Amy McGovern, Derek H. Rosendahl, and Rodger A. Brown

Abstract Tornadoes, which are one of the most feared natural phenomena, present a significant challenge to forecasters who strive to provide adequate warnings of the imminent danger. Forecasters recognize the general environmental conditions within which a tornadic thunderstorm, called a supercell thunderstorm, will form. They also recognize a supercell thunderstorm with its rotating updraft, or mesocyclone, when it appears on radar. However, only a minority of supercell storms produce tornadoes. Although most tornadoes are warned in advance, the majority of the tornado warnings are false alarms. In this chapter, we discuss the development of novel spatiotemporal data mining techniques for discriminating between supercell storms that produce tornadoes and those that do not. To test the novel techniques, we initially applied them to numerical models having coarse 500 meter horizontal grid spacing that did not resolve tornadoes but that did resolve the parent mesocyclones.

Keywords Spatiotemporal data mining • Tornado formation • Modeling • Random forests

A. McGovern (✉)
School of Computer Science, University of Oklahoma, 110 W Boyd St,
Norman, OK 73019, USA
e-mail: amcgovern@ou.edu

D.H. Rosendahl
School of Meteorology, University of Oklahoma, 120 David L. Boren Blvd. Suite 5900,
Norman, OK 73072, USA
e-mail: drose@ou.edu

R.A. Brown
National Severe Storms Laboratory/National Oceanic and Atmospheric Administration,
120 David L Boren Blvd, Norman, OK 73072, USA
e-mail: Rodger.Brown@noaa.gov

G. Cervone et al. (eds.), *Data Mining for Geoinformatics: Methods and Applications*,
DOI 10.1007/978-1-4614-7669-6_2, © Springer Science+Business Media New York 2014

1 Motivation

Tornadoes, which are one of the most feared natural phenomena, present a significant challenge to forecasters who strive to provide adequate warnings of the imminent danger. Forecasters recognize the general environmental conditions within which a tornadic thunderstorm, called a supercell thunderstorm, will form. They also recognize a supercell thunderstorm with its rotating updraft, or mesocyclone, when it appears on radar. However, only a minority of supercell storms produce tornadoes. There are no obvious clues within any of the routinely observed data to indicate which supercell storms are going to produce tornadoes and which ones are not. So to be on the safe side, forecasters issue a tornado warning whenever they detect on radar a supercell thunderstorm with a strengthening mesocyclone, which is the parent circulation within which tornadoes form. This approach results in the warning being issued an average 10–15 min before the appearance of a tornado, but a tornado appears only 20–30% of the time that a warning is issued, which results in a large percentage of false alarms (e.g., Simmons and Sutter 2011).

Surveys conducted by the National Weather Service following devastating U.S. tornadoes reveal that these false alarms are one of the factors contributing to desensitization on the part of the public concerning the need to adhere to warnings (e.g., NWS 2009, 2011). Having heard many warnings when a tornado did not form, members of the public tend to ignore the warning and only take shelter if they see a tornado approaching, with the result that some of them do not make it to a safe place in time. Therefore, there is a need to explore ways in which to identify unique factors within storms that lead to tornado formation.

One way to help understand the evolutionary characteristics of supercell thunderstorms, and especially those that produce tornadoes, is to conduct numerical modeling studies of the storms (e.g., Wicker and Wilhelmson 1995; Noda and Niino 2005; Xue et al. 2007). These types of studies typically investigate a single storm that develops within a given environment. A more informative approach is to conduct a number of fine-resolution (i.e., horizontal grid spacing less than 100 m) numerical modeling studies of supercell storms under a variety of environmental conditions. Data mining techniques then can be applied to the modeling results in order to discover the differences between supercell storms that produce tornadoes and those that do not.

In this chapter, we discuss the development of novel spatiotemporal data mining techniques that were initially applied to numerical models having coarse 500 m horizontal grid spacing that did not resolve tornadoes but did resolve the parent mesocyclones. Note that much of this chapter is derived from the following papers (Rosendahl 2008; Supinie et al. 2009; McGovern et al. 2010, 2011a,b, 2013). Although we are in the process of generating 75–100 numerically-modeled supercell thunderstorms using tornado-resolving 75 m horizontal grid spacing, we focus our discussion in this chapter on the 500 m resolution storms. We anticipate that the mining of our new higher-resolution data set will reveal more clues about the processes that lead to tornado formation within some supercell thunderstorms but not within others. This approach is discussed at the end of the chapter.

2 Natural Hazard Domain: Severe Storm Simulations

Rosendahl (2008) created a set of 261 simulations of supercell thunderstorms, which are the most severe type of thunderstorm and which generate the most violent tornadoes. Each simulation was generated using the Advanced Regional Prediction System (ARPS, Xue et al. 2000, 2001, 2003). The full details on the parameters chosen to create the storms are described in Rosendahl (2008). Each simulation is run for 3 h of storm time. The simulation saves the state of all relevant meteorological variables every 30 s of storm time for every grid point in the domain. With a resolution of 500 m horizontally and a domain size of 100 km by 100 km, a stretched vertical resolution focusing on the lower altitudes (with 50 voxels vertically) , the simulations must save over 100 different variables every 30 s for each of 2 million grid squares. In total, each simulation produces over 21 GB of data, which requires us to intelligently process and mine this data.

Although each simulation generates a full gridded field of meteorological variables, the variables near a storm cell will provide the most information. We identify and track storm cells using a modified form of the Storm Cell Identification and Tracking algorithm (Johnson et al. 1998; McGovern et al. 2007) where we track the cells based on their dominant updraft region (localized area with rising air) because it is the defining feature of a thunderstorm. Figure 1 shows an example of simulated radar reflectivity 90 min into a simulation. Reflectivity measures the

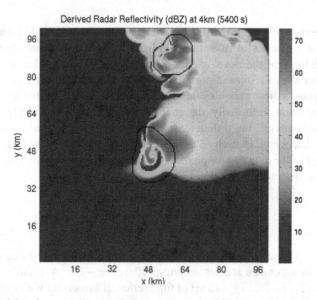

Fig. 1 Reflectivity of an example numerical storm simulation 90 min into the storm's lifetime. The scale on the right shows the intensity of the reflectivity in dBZ. Higher reflectivity regions indicate areas where the storm is producing intense precipitation. The *black outlines* highlight individual storm cells which are used to extract the storm metadata

intensity of the precipitation within the storm which means that regions with more intense rain, snow, ice, or hail have higher reflectivity values. The black outlines in Fig. 1 show the two storm regions that are being tracked during that period. Because weak short-lived storms are not of interest in this study, we only track cells that last for at least 30 min. Each simulation typically produces 3–4 such cells. The 261 storm simulations generated 1,168 separate storm cells that each lasted at least 30 min.

3 Spatiotemporal Data Mining

Weather phenomena vary as a function of both time and space. It is difficult to ignore one aspect in favor of the other so our goal has been to develop spatiotemporal data mining algorithms. This chapter first reviews our time series approach and then presents several spatiotemporal algorithms that we have developed. Each storm defines a four dimensional region of interest for data mining. The first three dimensions are the spatial dimensions and the fourth dimension is time. Given the sheer size of the data, all of our approaches reduce the data by extracting high-level metadata and then mining the metadata.

3.1 Multi-variate Time Series Approach

In the first approach to improving our understanding about tornado formation, we seek to identify a series of rules that highlight how environmental characteristics must change to favor the development of tornadoes. We do not know in advance what characteristics are most important so the learning and mining algorithms must identify the most salient variables and discover the temporal motifs most predictive of tornadic rotation. Informally, the goal of our approach is to identify the most relevant dimensions of a multi-dimensional time series, grow a set of predictive rules from motifs discovered in each of those dimensions, and use these to improve our understanding of the data and for prediction. We briefly review definitions necessary for our time series mining algorithm.

Definitions

Definition 1. A time series $T = \langle t_1, t_2, \ldots, t_{n-1}, t_n \rangle$ is an ordered sequence of real-valued observations taken at discrete times: $1, 2, \ldots, n-1, n$. A d-dimensional time series $T^d = \langle T_1, T_2, \ldots, T_d \rangle$ is a set of time series all associated with a single event and correlated in time.

This is the standard definition of time series (e.g., Mueen et al. 2009). Our examples are all temporally ordered sequences but other orderings are possible.

Rather than examining only a single attribute as it varies in time, we assume that the event can be measured with a variety of attributes (d in this definition), each of which is measured on the same discrete time interval. These measurements are not required to be independent of one another. Our severe weather simulations have $d = 100$ but the independent dimensionality of the data is much less (approximately $d = 40$).

Definition 2. A labeled multi-dimensional time series is a tuple $E = \{T^d, l\}$ where T^d is a d-dimensional time series and $l \in \mathscr{L}$ where $\mathscr{L}$ is a discrete set of labels (and it is not required to be binary).

Because each of our d-dimensional time series is associated with an event, each example E_i is labeled. In the case of our severe weather data, $\mathscr{L}$ is either binary (positive/negative) or takes on one of three possible values: positive, negative, or intermediate. The approach does not restrict the possible numbers of labels but it does require that the cardinality of $\mathscr{L}$ be finite.

Our data set, $D = \langle E_1, E_2, \ldots, E_n \rangle$, consists of a set of labeled multi-dimensional time series, each of which can last for a variable amount of time but each of which is assumed to have the same dimensionality. That is, all attributes that measure an event are assumed to be present in each labeled example.

Definition 3. A single dimensional time series motif $M_j = \langle t_i, t_{i+1}, \ldots, t_{i+m} \rangle$ consists of a temporally ordered subsequence of a time series where $1 \leq i \leq n$ and $0 < m \leq n$. This motif is of length m and is on dimension j where $1 \leq j \leq d$.

As stated above, our goal is to identify multi-dimensional times series motifs that can be used for prediction. As such, we build on the previous definition.

Definition 4. A multi-dimensional time series motif $P = \langle M_{i_1}, M_{i_2}, \ldots M_{i_m} \rangle$ is a temporally ordered set of single-dimensional time series motifs (see Definition 3). The temporal ordering specifies that the initiation of each single-dimensional time series M_{i_j} must begin after or simultaneously with the previous single-dimensional time series in the set $M_{i_{j-1}}$.

A multi-dimensional time series motif does not specify how many dimensions of the overall available dimensions must be used and it can even repeat dimensions, given that each is temporally ordered. For example, the first motif may be on dimension 1, the second on dimension 3, and the third on dimension 1 again. The temporal ordering is not strict as it requires that M_j begins after M_{j-1} but simultaneous initiations are also acceptable. M_j cannot begin before M_{j-1}. This definition differs slightly from the definition in Mueen et al. (2009), where there is a requirement for a temporal gap in between the different subsequences. We were interested in rules that could identify two features both firing at once and did not impose the temporal gap.

Algorithm 1: Grow multi-dimensional time series motifs/rules

Input: D: training data, SAX parameters (alphabet size, word size, averaging interval),
        minimum POD, maximum FAR
Output: A list of rules sorted by CSI
foreach *dimension d* **do**
  | Discretize E_i^d for all examples i using SAX
  | Build trie with pointers to the start and to the end of each word in the sliding window
end
foreach *dimension d* **do**
  | Identify all single dimensional words with minimum POD and maximum FAR
  | Recursively grow longer rules within dimension d
end
for *all rules that meet minimum POD and maximum FAR criteria* **do**
  | Grow rules across dimensions
end
return *list of rules sorted by CSI score*

Algorithm

A general outline of our approach for identifying multi-dimensional time series motifs is given in Algorithm 1 and we describe each step in detail below. The general idea is to search for the critical dimensions of the data by identifying single dimensional motifs first, narrow down the set of possible single dimensional motifs using user specified minimum performance metrics and then grow the motifs across dimensions using the single dimensional motifs as building blocks for larger motifs.

The rules are built and scored using three standard performance measures. The probability of detection (POD) is the number of times that an event was correctly predicted divided by the total number of observed events. POD ranges from 0 to 1 with 1 representing a perfect score. The false alarm ratio (FAR) is the number of times that an event was incorrectly predicted to occur divided by the total number of events predicted to occur. FAR ranges from 0 to 1 with 0 representing a perfect score. The critical success index (CSI, Donaldson Jr et al. 1975; Schaefer 1990) evaluates success as a function of only the events that are predicted to be positive and the ones that were actually positive. Thus, CSI ignores the true negatives and provides an index that incorporates both POD and FAR into one measure. It ranges from 0 to 1 with 1 being a perfect score (perfect POD and perfect FAR). This measure is particularly important for assessing rare events such as tornadoes where POD and FAR alone are inadequate measures of predictability. For instance, a perfect score can be attained for POD by simply predicting an event (e.g., tornado) to occur every time or for FAR by never predicting an event to occur. CSI, however, combines both of these and therefore gives a more robust performance measure.

Because a brute force approach to searching multi-dimensional data grows exponentially in the number of dimensions and the size of the data, it quickly becomes intractable. Some researchers approach this issue using a random type of search (Minnen et al. 2007; Vahdatpour et al. 2009) but our pruning and discretization enables us to search the full space efficiently. Additionally, searching for motifs in real-valued data is difficult, as has been noted by many researchers (for example, Das et al. 1998; Chiu et al. 2003; Mueen et al. 2009). We chose to address this latter problem using the SAX discretization technique (Lin et al. 2003) and we address the computational aspects using a combination of approximate search and intelligent data structures such as the trie described in Keogh et al. (2005).

The first step of our approach is to discretize each of the dimensions of data using SAX, which is a standard time series discretization technique (e.g., Lin et al. 2003; Shieh and Keogh 2009; Keogh et al. 2005; Lin et al. 2007; Minnen et al. 2007; Vahdatpour et al. 2009). To do this, we discretize all examples at once for each dimension. This ensures that the discretizations can be easily compared and that a in one time series has a similar meaning to a in another example. Given the number of dimensions and examples, multiple passes through the data would be computationally expensive. To address this, we make use of the trie data structure as discussed in Keogh et al. (2005). Since each leaf in the trie has information about exactly which time series that word occurs in, the trie also stores the POD and FAR measures for use in the mining. We have demonstrated in previous work that the results are not sensitive to the choices of parameters to SAX (McGovern et al. 2011b).

Once the tries are built, the data mining algorithm makes use of them to efficiently narrow down the search for more complicated motifs. To do this, we look through each dimension of the data and narrow down the set of basic SAX words using user specified thresholds of the POD and FAR performance measures. By specifying a minimum POD and a maximum FAR, we limit the number of elementary motifs identified. These numbers can come from a user's experience in the domain. This significantly improves the running time of the search since all possible combinations of small motifs are used to grow the larger motifs. Since the search proceeds from general motifs (e.g. short ones) to specific motifs (longer multi-dimensional motifs), the performance of the motifs will only improve POD and FAR. This is similar to the pruning search discussed in Webb (1995), Oates and Cohen (1996), and McGovern and Jensen (2008) where the type of search and evaluation measures can be combined to enable admissible pruning. Thus, we specify minimum levels of performance that would be acceptable and expect that the final numbers will be significantly better with the more specific rules.

Once the basic words are identified, we grow the motifs recursively using the basic words that pass the POD/FAR thresholds. The recursive growth of the motifs within a dimension works by doubling each motif, similar to the method employed by SPADE (Zaki 2001). Each motif is doubled while maintaining the minimum POD and maximum FAR requirements. For example, $[a, b]$ can be doubled to $[a, b, a, b]$ and it can also be combined with other valid words such as $[a, b, b, a]$. When a doubling fails, the motif is grown linearly by adding words until it is at its maximum length. Since motifs are grown in chunks of word size, not all possible motifs can be detected. For example, the motif $[a, b, a]$ is not discoverable in the example because the minimum word size is 2. However, in all of our experiments, we keep the word size small to minimize this issue.

The motif growing within a single dimension is repeated for all dimensions before searching across the dimensions. Although this growing sounds computationally expensive, only one pass through the original data is required. Using the trie data structure, all further motifs can be grown by examining the starting and ending times of each individual motif and ensuring that they follow one another temporally (e.g. they satisfy Definition 4 above). The POD and FAR measures continue to be directly computed from the trie by intersecting the positive and negative graphs computed for each individual piece of the motif.

The last step of the algorithm is to repeat the search by combining the different words across the dimensions of the data. Once all of the motifs have been identified for each dimension, an exhaustive search of temporal orderings across dimensions is performed. Although doing an exhaustive search sounds infeasible, it is possible because of the admissible pruning afforded by the user's minimum POD and maximum FAR measures. In addition, the O(1) access time of the trie facilitates the overall approach. Without the pruning, this approach would be untenable but the pruning significantly improves the running time by enabling the search to ignore large portions of the search space while guaranteeing that the rules that could be identified in that space will never meet the user's specified performance measures. In addition, the trie again can be used to quickly compute the POD/FAR measures across dimensions by continuing to intersect the positive and negative series observed by each piece of the motif.

Empirical Results

To create the metadata for the time series data mining, we extract maximum and minimum values of relevant meteorological quantities within each of the simulated storm cells. We measure the maximum and minimum value for each variable from the surface to 2 km in height and then from 2 to 8 km. For some variables, we also store the maximum and minimum values at the surface. This allows us to identify whether a maximum or minimum value is associated with a surface, low, or mid to upper altitude feature. This yields a 100 dimensional time series for each storm. The full set of quantities is defined in Rosendahl (2008).

We extract the maximum and minimum values every 30 s for the entire 3 h of simulation. Figure 2 shows an example of several time series extracted for two of the meteorological quantities. The left panel shows the evolution of the vertical vorticity (an instantaneous measure of spin about a vertical axis) at the surface, low altitudes, and mid- to upper-altitudes for the center storm shown in Fig. 1. Positive (maximum) and negative (minimum) values in the left panel correspond to counterclockwise and clockwise rotation respectively. The right panel shows the maximum horizontal wind speed values.

Given the sheer number of storm cells, we developed an automated labeling approach based on the key characteristics of tornadic storms. Because the horizontal grid spacing of the simulations is too coarse to detect rotation on the scale of a tornado and creating higher resolution simulations requires exponentially more

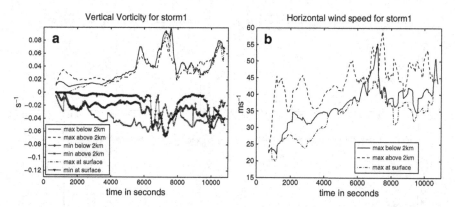

Fig. 2 Meteorological quantities extracted from an example storm. Shown are the maximum and minimum quantities for vertical vorticity (*left*) and the maximum values for horizontal wind speed (*right*). (**a**) Example vertical vorticity data. (**b**) Example horizontal wind speed data

computational time and space, we labeled each storm as to whether it produced strong low-altitude rotation ("positive"), produced either no or very weak rotation ("negative") or was in between these two categories ("intermediate" or "maybe"). Strong low-altitude rotation was defined as a storm where there was a decrease in the surface pressure perturbation of at least -900 Pa in 1,000 s and either an increase in the horizontal wind speed at the surface of at least $5\,\mathrm{m\,s^{-1}}$ within 750 s or an increase in the absolute value of the vertical vorticity of at least $0.03\,\mathrm{s^{-1}}$ within 500 s. These features had to overlap within a 600 s window to ensure that they were correlated. Storms where the pressure drop fell within the range of -900 to -300 Pa and met the vertical vorticity and wind speed criteria or that had a pressure drop but no corresponding increase in vertical vorticity or wind speed were labeled as intermediate storms. The remainder of the storms were labeled as negative storms. This yielded 58 positive storms, 373 intermediate storms, and 737 negatives.

Given the labeled data, we further processed it in two ways. First, if we were to feed all of the time series information to the data mining algorithm, it would identify the approach that we took to label the data. To avoid rules that simply state that pressure perturbations or vertical vorticity are critical, we remove from consideration each of the features used to label the data from consideration. Further, to ensure that identified precursors were associated with the developing strong low-altitude rotation in positive storms, we saved data for 30 min immediately prior to the beginning of the corresponding pressure drop, which was defined using a Gaussian derivative filter on the pressure perturbation time series. For non-positive cases, we randomly sampled 30 min to avoid obvious labeling based on the length of the time series given to the data mining algorithm.

Figure 3 shows an example rule identified by the time series data mining algorithm. We explored a wide variety of parameter variations and show only a single example rule, due to space considerations. This rule identifies four salient environmental characteristics associated with rotation in our simulations and also

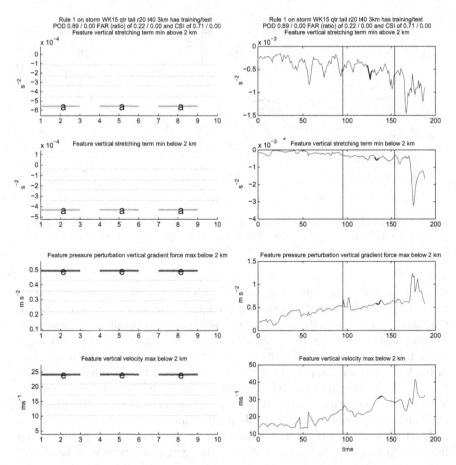

Fig. 3 Example rule identified by time series mining algorithm. The *left column* identifies 3-letter words comprising the rule in sequential order (baroclinic generation term (vertical) min below 2 km, vertical stretching term max below 2 km, pressure perturbation vertical gradient force max above 2 km and vertical stretching term min below 2 km). Time step within each word plot in *left column* is arbitrary but colored line letter segments correspond to a 30 s time period (one output interval). The five equiprobable Gaussian regions associated with each word are demarcated by *light gray* horizontal lines. *Right column* provides meteorological quantity metadata from an example storm that contains the rule. Each word from the *left column* is identified by a *black line* segment in the metadata. The 30 min window prior to the development of strong low-level rotation is contained within the two vertical *gray lines*. Corresponding performance measures are listed at the *top* of the figure

shows the behaviors of each measurement. In general, the rule indicates that strong low-altitude rotation has a greater probability of developing when the four identified meteorological quantities reach relatively extreme values in a short time span. Hundreds of such rules were identified by the data mining algorithm and were ranked according to their performance measure scores.

This information can be highly useful to those in the meteorological community because each rule is simply a sequence of events taking place within a storm and therefore identifying the most important of these sequences may offer insight into how tornadoes form and how one can better predict their occurrence. Also, the automated method we used provides an innovative alternative for assessing storm feature evolution which otherwise would require a brute force approach of evaluating individual sequences of model output across hundreds of simulated storms.

3.2 Spatiotemporal Relational Probability Trees

The multi-variate time series analysis was promising and demonstrated that mining of the storm data would likely yield useful information. However, our long-term goal was to develop an approach that would reason with both the spatial and the temporal nature of the data. As such, we developed the Spatiotemporal Relational Probability Tree (SRPT). The SRPT is a probability estimation tree that learns with spatiotemporally varying relational data. For example, a SRPT can predict new class labels based on questions such as "Has a downdraft lasted for at least 5 min?" or "Did a region of strong tilting of horizontal vorticity appear before the downdraft doubled in intensity?" While the SRPT was inspired by the relational probability tree (RPT) (Neville et al. 2003), the SRPT represents both the data and the decision tree distinctions in a very different manner.

Data Representation

By moving to a relational representation of our weather data, we gain the ability to reason about the high-level objects already identified by meteorologists. Such objects can represent concepts that they believe are associated with rotations and tornadoes or even different regions of a storm cell. The relational representation enables us to reason about the spatial or spatiotemporal relationships between the objects. Our data are represented as spatiotemporal attributed relational graphs, as we first presented in McGovern et al. (2008). This representation is an extension of the attributed graph approach to handle spatiotemporally varying data (Neville et al. 2003; Neville and Jensen 2004; Jensen 2005). All *objects*, such as updrafts or hail cores, are represented by vertices in the graph. *Relationships* between the objects are represented using edges. With the severe weather data, the majority of the relationships are spatial. Both objects and relationships can have *attributes* associated with them and these attributes can vary both spatially and temporally. In the case of a spatially or spatiotemporally varying attribute, the data are represented as either a scalar or a vector field, depending on the nature of the data. The ability to represent spatial fields of objects is a key addition to the SRRF results presented here. This field can be two or three dimensional in space and can also vary as a

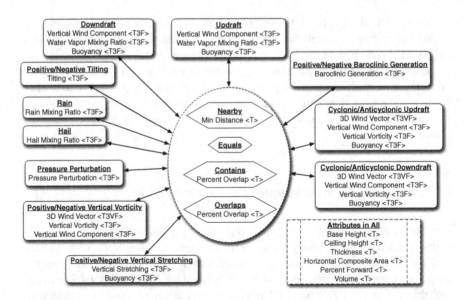

Fig. 4 Schema for the simulated supercell storms domain. The *rounded rectangles* represent objects and the *hexagons* show the relations. The type of each object or relation is *bolded* and the *arrows* show the directionality of the relationships. Attributes are listed inside both objects and relations. *T* denotes a temporally varying attribute; *T3F* denotes a three-dimensional spatiotemporal field

function of time. In addition to attributes varying over space and time, the existence of objects and relationships can also vary as a function of time. If an object or a relationship is *dynamic*, it has a starting and an ending time associated with it.

Figure 4 shows a schema of our spatiotemporal relational data for the severe storms domain. Note that this shows the possible objects and relationships within a single example graph but it does not show the specific instantiation of a graph as that will differ for each storm. Both objects and relations are required to be assigned a type and graphs are required to be labeled (the label does not have to be binary). Objects and relations can be either static (existing for the duration of the graph) or dynamic. Temporal consistency is enforced for all dynamic objects or relations. Attributes can be associated with both objects and relations. Attributes can also be static or dynamic. If an attribute is static, its value stays the same throughout the lifetime of the object or relation. If it is dynamic, it can vary as a function of time or with space and time.

Algorithm

The SRPT is a decision-type tree with the ability to differentiate the data based on spatial, temporal, and spatiotemporal questions at each node. The tree is learned in the standard greedy manner and the differences lie in the questions the SRPT can ask at each node. In a standard decision tree such as C4.5 (Quinlan 1993), there are a finite number of possible questions about the data. Because there are a very large number of possible splits or questions for spatiotemporal data, we sample the specific splits using a user specified sampling rate. For each sample, a split template is selected randomly and the pieces of the template are filled in using randomly chosen examples in the training data. These templates are described below.

The non-temporal splits are:

- Exists: Does an object or relation of a particular type exist?
- Attribute: Does an object or a relation with attribute a have a [MAX, MIN, AVG, ANY] value $\geq$ than a particular value v?
- Count Conjugate: Are there at least n yes answers to distinction d? Distinction d can be any distinction other than Count Conjugate.
- Structural Conjugate: Is the answer to distinction d related to an object of type t through a relation of type r? Distinction d can be any distinction other than Structural Conjugate.

The temporal splits are:

- Temporal Exists: Does an object or a relation of a particular type exist for time period t?
- Temporal Ordering: Do the matching items from basic distinction a occur in a temporal relationship with the matching items from basic distinction b? The seven types of temporal ordering are: *before, meets, overlaps, equals, starts, finishes,* and *during* (Allen 1991).
- Temporal Partial Derivative: Is the partial derivative with respect to time on attribute a on an object or relation of type $t \geq v$?

The spatial and spatiotemporal splits are:

- Spatial Partial Derivative: Is the partial derivative with respect to space of attribute a on an object or relation of type $t \geq v$?
- Spatial Curl: Is the curl of fielded attribute $a \geq v$?
- Spatial Gradient: Is the magnitude of the gradient of fielded attribute $a \geq v$?
- Shape: Is the primary 3D shape of a fielded object a cube, sphere, cylinder, or cone? This question also works for 2D objects and uses the corresponding 2D shapes.
- Shape Change: Has the shape of an object changed from one of the primary shapes over to a new shape over the course of t steps?

Empirical Results

Figure 5 shows an example SRPT learned from the severe storm simulation data. Note that the high probabilities of the formation of strong low-altitude rotation occur only along the left portion of the tree. The low probabilities of low-altitude rotation occur along the right portion of the tree. The category "maybe" is found throughout the tree.

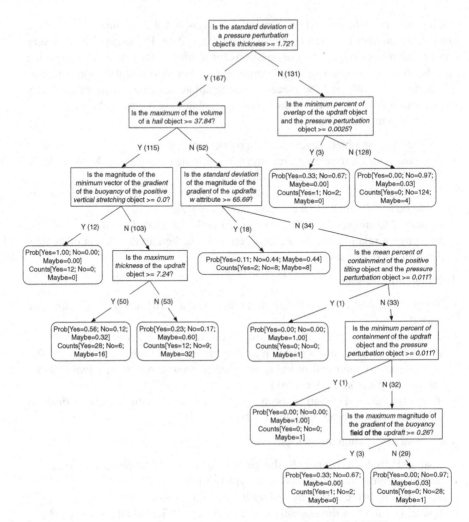

Fig. 5 Highest scoring single tree in the simulated storms domain using the SRPT

3.3   Spatiotemporal Relational Random Forests

Although one reason for developing a tree-based model was for its human readability by domain scientists, a single decision-tree type model is known to be brittle (Pérez et al. 2005; Dwyer and Holte 2007). This brittleness makes it harder for a domain scientist to trust the results from a single tree. As such, we developed an ensemble approach of SRPTs, following the Random Forest paradigm (Breiman 2001). The Spatiotemporal Relational Random Forest (SRRF) enables a domain scientist to grow a robust predictive model of spatiotemporally varying relational data. Similar to Random Forests, the SRRF provides a method for variable importance, which measures the importance of attributes on objects or relationships to the predictive power of the forest.

Method

Growing a SRRF is very similar to the approach used to grow a Random Forest with the only differences occurring in the individual tree growing algorithm (described above) and from the nature of the spatiotemporal relational data. The training data for each tree in the forest is created using a bootstrap resampling of the original training data. The difference in the learning methods arises from the nature of the spatiotemporal relational data and the SRPTs versus C4.5 trees. In the RF algorithm, each node of each tree in the forest was trained on a different subset of the available attributes. Since the individual trees were standard C4.5 decision trees, this limited the number of possible splits each tree could make. Because each tree was also trained on a different bootstrap resampled set of the original data, the trees were sufficiently different from one another to make a powerful ensemble. Because there are a very large number of possible splits that the SRPTs can choose from, an SRPT finds the best split through sampling, as described above. Like the original RF trees, SRPTs are still built using the best split identified at each level. With fewer samples, these splits may not be the overall best for a single tree, but they will be sufficiently different across the sets of trees to ensure diversity in the forest. However, if the number of samples is too small, the number of trees needed in the ensemble to obtain good results may be prohibitively large.

For a particular attribute a, RFs measure variable importance by querying each tree in the forest for its vote on the out-of-bag data. Then, the attribute values for attribute a are permuted within the out-of-bag instances and each tree is re-queried for its vote on the permuted out-of-bag data. The average difference between the votes on the unpermuted data for the correct class and the votes for the correct class on the permuted data is the raw variable importance score. We have directly converted this approach to the SRRFs and can measure variable importance on any attribute of an object or relation. Spatially and temporally varying attributes are treated as a single entity and permuted across the objects/relations but their spatial and/or temporal ordering is preserved. We examine the variable importance in each of our data sets.

Empirical Results

Table 1 shows the top 10 most statistically significant attributes associated with storms that develop strong low-altitude rotation, measured using variable importance on the SRRFs. Because rotation is associated with pressure drops, the inclusion of multiple attributes of the pressure perturbation object seems quite reasonable. Updrafts are the thermodynamic engines that power supercell thunderstorms and

Table 1 Top 10 statistically significant attributes according to variable importance on the simulated storms data

Fields	Mean variable importance
PressurePerturbation.PressurePerturbationField	0.251
PressurePerturbation→ Overlap.PercentOverlap→ Rain	0.188
Hail.HorizontalCompositeArea	0.114
PressurePerturbation.Volume	0.107
PressurePerturbation.Thickness	0.101
PressurePerturbation.BaseHeight	0.099
Hail.Volume	0.089
Updraft.Volume	0.086
Updraft.Thickness	0.061
Updraft.Buoyancy	0.060

they play a major role in the concentration of rotation into tornado-scale vortices. Owing to the presence of strong updrafts within such storms, conditions also are favorable for the formation of hail. Therefore, it is logical that the SRRF approach should find that pressure, updrafts, and hail attributes are important features of supercell storms that develop significant low-altitude rotation.

4 Ongoing Research

The methods described in this chapter have proven both promising and fruitful in mining the 500 m resolution storm simulations. However, these simulations are limited in that they cannot resolve the circulation within a tornado. In our current research, we are developing a set of high resolution simulations, capable of resolving tornadoes. These simulations have horizontal grid spacings of 75 m. Figure 6 shows the simulated reflectivity from one of these storms. This storm generated a tornado approximately 2 h into the simulation and the effects of the tornado can be seen on the reflectivity.

We are also developing several new spatiotemporal relational data mining methods. The first focuses on enhancements to the SRPT/SRRF approach that will enable the true discovery of spatial and temporal relationships, not pre-specified by the user. The second focuses on using Bayesian Network structure learning techniques to identify salient relationships in the data.

Acknowledgements This material is based upon work supported by the National Science Foundation under IIS/CAREER/0746816 and corresponding REU Supplements IIS/0840956, 0938138, 1036023, 1129292, and the NSF ERC Center for Collaborative Adaptive Sensing of the Atmosphere (CASA, NSF ERC 0313747).

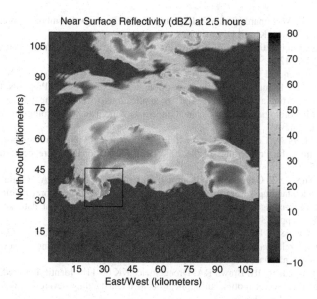

Fig. 6 Simulated near-surface reflectivity of a 75 m horizontal resolution storm. The *boxed* region highlights a tightly wound up end of the hook echo that indicates the presence of a tornado

References

Allen JF (1991) Time and time again: the many ways to represent time. Int J Intell Syst 6(4): 341–355

Breiman L (2001) Random forests. Mach Learn 45(1):5–32

Chiu B, Keogh E, Lonardi S (2003) Probabilistic discovery of time series motifs. In: In the 9th ACM SIGKDD international conference on knowledge discovery and data mining, Washington, DC, pp 493–498

Das G, Lin K, Mannila H, Renganathan G, Smyth P (1998) Rule discovery from time series. In: Proceedings of the ACM SIGKDD international conference on knowledge discovery and data mining, New York, pp 16–22

Donaldson Jr RJ, Dyer RM, Kraus MJ (1975) An objective evaluator of techniques for predicting severe weather events. In: Preprints: ninth conference on severe local storms, American Meteorological Society, Norman, pp 321–326

Dwyer K, Holte R (2007) Decision tree instability and active learning. In: ECML '07: proceedings of the 18th European conference on machine learning, Warsaw. Springer-Verlag, Berlin, pp 128–139

Jensen D (2005) Proximity knowledge discovery system. http://kdl.cs.umass.edu/proximity

Johnson JT, MacKeen PL, Witt A, Mitchell ED, Stumpf GJ, Eilts MD, Thomas KW (1998) The storm cell identification and tracking algorithm: an enhanced WSR-88D algorithm. Weather Forecast 13(2):263–276

Keogh E, Lin J, Fu A (2005) HOT SAX: efficiently finding the most unusual time series subsequence. In: Proceedings of the 5th IEEE international conference on data mining (ICDM 2005), Houston, pp 226–233

Lin J, Keogh E, Lonardi S, Chiu B (2003) A symbolic representation of time series, with implications for streaming algorithms. In: Proceedings of the 8th ACM SIGMOD workshop on research issues in data mining and knowledge discovery, San Diego, pp 2–11

Lin J, Keogh E, Li W, Lonardi S (2007) Experiencing SAX: a novel symbolic representation of time series. Data Min Knowl Discov 15(2):107–144

McGovern A, Jensen D (2008) Optimistic pruning for multiple instance learning. Pattern Recognit Lett 29(9):1252–1260

McGovern A, Rosendahl DH, Kruger A, Beaton MG, Brown RA, Droegemeier KK (2007) Anticipating the formation of tornadoes through data mining. In: Preprints of the fifth conference on artificial intelligence and its applications to environmental sciences at the American Meteorological Society annual meeting, American Meteorological Society, San Antonio, Paper 4.3A

McGovern A, Hiers N, Collier M, Gagne II DJ, Brown RA (2008) Spatiotemporal relational probability trees. In: Proceedings of the 2008 IEEE international conference on data mining, Pisa, pp 935–940

McGovern A, Supinie T, Gagne II DJ, Troutman N, Collier M, Brown RA, Basara J, Williams J (2010) Understanding severe weather processes through spatiotemporal relational random forests. In: Proceedings of the 2010 NASA conference on intelligent data understanding, Mountain View, pp 213–227

McGovern A, Gagne II DJ, Troutman N, Brown RA, Basara J, Williams J (2011a) Using spatiotemporal relational random forests to improve our understanding of severe weather processes. Stat Anal Data Min 4(4):407–429

McGovern A, Rosendahl DH, Brown RA, Droegemeier KK (2011b) Identifying predictive multi-dimensional time series motifs: an application to understanding severe weather. Data Min Knowl Discov 22(1):232–258

McGovern A, Troutman N, Brown RA, Williams JK, Abernethy J (2013) Enhanced spatiotemporal relational probability trees and forests. Data Min Knowl Discov 26(2):398–433

Minnen D, Isbell C, Essa I, Starner T (2007) Detecting subdimensional motifs: an efficient algorithm for generalized multivariate pattern discovery. In: Proceedings of the 2007 seventh IEEE international conference on data mining (ICDM '07), Omaha, pp 601–606

Mueen A, Keogh E, Zhu Q, Cash S, Westover B (2009) Exact discovery of time series motifs. In: Proceedings of the SIAM international conference on data mining, Sparks, pp 473–484

Neville J, Jensen D (2004) Dependency networks for relational data. In: Proceedings of the fourth IEEE international conference on data mining, Brighton, pp 170–177

Neville J, Jensen D, Friedland L, Hay M (2003) Learning relational probability trees. In: Proceedings of the ninth ACM SIGKDD international conference on knowledge discovery and data mining. ACM, New York, pp 625–630

Noda A, Niino H (2005) Genesis and structure of a major tornado in a numerically–simulated supercell storm: importance of vertical vorticity in a gust front. Sci Online Lett Atmos 1:5–8

NWS (2009) Service assessment, Mother's Day weekend tornado in Oklahoma and Missouri, May 10, 2008. http://www.nws.noaa.gov/os/assessments/pdfs/mothers_day09.pdf

NWS (2011) Nws central region service assessment, Joplin, Missouri, tornado – May 22, 2011. http://www.nws.noaa.gov/os/assessments/pdfs/Joplin_tornado.pdf

Oates T, Cohen PR (1996) Searching for structure in multiple streams of data. In: Proceedings of the thirteenth international conference on machine learning. Morgan Kaufmann, Bari, pp 346–354

Pérez JM, Muguerza J, Arbelaitz O, Gurrutxaga I, Martìn JI (2005) Consolidated trees: classifiers with stable explanation. A model to achieve the desired stability in explanation. In: Lecture notes in computer science, Springer Berlin, Heidelberg, pp 99–17

Quinlan JR (1993) C4.5: programs for Machine Learning. Morgan Kaufmann, San Francisco

Rosendahl DH (2008) Identifying precursors to strong low-level rotation within numerically simulated supercell thunderstorms: a data mining approach. Master's thesis, School of Meteorology, University of Oklahoma

Schaefer JT (1990) The critical success index as an indicator of warning skill. Weather Forecast 5(4):570–575

Shieh J, Keogh E (2009) iSAX: indexing and mining terabyte sized time series. In: Proceedings of the IEEE international conference on data mining

Simmons KM, Sutter D (2011) Economic and societal impacts of tornadoes. American Meteorological Society, Boston

Supinie T, McGovern A, Williams J, Abernethy J (2009) Spatiotemporal relational random forests. In: Proceedings of the IEEE international conference on data mining (ICDM) workshop on spatiotemporal data mining .p electronically published

Vahdatpour A, Amini N, Sarrafzadeh M (2009) Toward unsupervised activity discovery using multi-dimensional motif detection in time series. In: Proceedings of the 21st international joint conference on artificial intelligence (IJCAI'09), Pasadena, pp 1261–1266

Webb GI (1995) OPUS: an efficient admissible algorithm for unordered search. J Artif Intell Res 3:431–465

Wicker LJ, Wilhelmson RB (1995) Simulation and analysis of tornado development and decay within a three–dimensional supercell thunderstorm. J Atmos Sci 52(15):2675–2703

Xue M, Droegemeier KK, Wong V (2000) The Advanced Regional Prediction System (ARPS) - a multiscale nonhydrostatic atmospheric simulation and prediction model. Part I: model dynamics and verification. Meteorol Atmos Phys 75:161–193

Xue M, Droegemeier KK, Wong V, Shapiro A, Brewster K, Carr F, Weber D, Liu Y, Wang D (2001) The Advanced Regional Prediction System (ARPS) – a multiscale nonhydrostatic atmospheric simulation and prediction tool. Part II: model physics and applications. Meteorol Atmos Phys 76:143–165

Xue M, Wang D, Gao J, Brewster K, Droegemeier KK (2003) The Advanced Regional Prediction System (ARPS), storm-scale numerical weather prediction and data assimilation. Meteorol Atmos Phys 82:139–170

Xue M, Droegemeier KK, Weber D (2007) Numerical prediction of high-impact local weather: a driver for petascale computing. In: Petascale computing: algorithms and applications. Chapman and Hall/CRC, Boca Raton, chap 18, pp 103–125

Zaki MJ (2001) Spade: an efficient algorithm for mining frequent sequences. Mach Learn 42(1/2):31–60. special issue on unsupervised learning

Source Term Estimation for the 2011 Fukushima Nuclear Accident

Guido Cervone and Pasquale Franzese

Abstract A new methodology is presented for the reconstruction of an unsteady release rate of an atmospheric contaminant, based on atmospheric transport and dispersion models, concentration measurements, and stochastic search techniques. The methodology is applied to reconstruct the radiation release rate for the March 2011 Fukushima nuclear accident.

The observed radiation data were retrieved from 218 stations located in 17 Japanese prefectures. The dispersion simulations are performed using the SCIPUFF model, using model vertical profiles and ground meteorological data. The non-stationary time-series of the Fukushima release rate is determined for a period of 5 days with a 2-h resolution.

Keywords Fukushima nuclear accident • Release rate estimation • Spatio-temporal optimization • Evolutionary algorithms • Dispersion modeling

1 Introduction

On 11 March 2011 at 05:46 UTC (14:46 local time, UTC +9) a massive Mw 9.0 underwater earthquake occurred 70 km offshore of the eastern coast of Japan, with epicenter at 38.322N and 142.369E. The earthquake generated a tsunami that rapidly

G. Cervone (✉)
Department of Geography and Institute for CyberScience, The Pennsylvania State University, 302 Walker Building, University Park, PA 16802, USA

Research Application Laboratory, National Center for Atmospheric Research, Boulder, CO 80307, USA
e-mail: cervone@ucar.edu

P. Franzese
Ecology and Environment, Inc., 368 Pleasant View Dr, Lancaster, NY 14086, USA
e-mail: pfranzese@ene.com

G. Cervone et al. (eds.), *Data Mining for Geoinformatics: Methods and Applications*,
DOI 10.1007/978-1-4614-7669-6_3, © Springer Science+Business Media New York 2014

hit the eastern coast of Japan, and propagated across the Pacific ocean to the western coast of the Americas.[1] The tsunami wave hit the Fukushima power plant about 40 min after the earthquake, leading to the catastrophic failure of the cooling system.

Several radioactive releases ensued as a result of an increase of pressure and temperature in the nuclear reactor buildings. Some releases were the result of both controlled and uncontrolled venting, while others were the result of the explosions that compromised the containment structures. The explosions were most likely caused by ignited hydrogen, generated by zirconium-water reaction occurring after the reactor core damage.

The largest radioactive leaks occurred between the 12 and 21 March 2011. Radioactivity was recorded at different locations throughout Japan on the ground, in the water and in the air. The individual radionuclide distributions assessed by Kinoshita et al. (2011) over central-east Japan from the Fukushima nuclear accident indicate that the prefectures of Fukushima, Ibaraki, Tochigi, Saitama, and Chiba and the city of Tokyo were contaminated by doses of radiations, and that a large amount of radioactivity was discharged on March 15th and 21st.

The radioactive cloud was quickly transported around the world, reaching within a few days North America and Europe (Potiriadis et al. 2012; Masson et al. 2011). Radioactive concentrations were recorded along the US West Coast (Bowyer et al. 2011). Estimating the fate of the contaminants and predicting their health impact quickly became an issue of great importance (Calabrese 2011).

Transport and dispersion (T&D) models can be used to compute atmospheric radioactivity and ground deposition. Various models are available depending on the scale of the problem. For instance, the exclusion zone and its evacuation are determined by dispersion simulations in the area immediately surrounding the nuclear reactor on a scale ranging from meters to a few kilometers. Contamination at a planetary scale can be assessed by long range transport models.

Dispersion simulations require meteorological data, terrain characteristics, source location and release rate. A major problem with the simulation of the Fukushima accident is the large uncertainty associated with the time-dependent release rate of radioactive contaminant. Yasunari et al. (2011) used the FLEXPART Lagrangian T&D dispersion model with a constant source term, and compared their results with the estimated total Caesium-137 deposition obtained by integrating daily observations in each prefecture in Japan. The results indicate heavy soil contamination by Caesium-137 in large areas of eastern and northeastern Japan, whereas western Japan was sheltered by mountain ranges.

In general, in order to estimate the release rate in case of a nuclear accident three approaches can be followed, based on:

1. Observations;
2. Nuclear reactor modeling;
3. Atmospheric modeling.

---

[1]http://ptwc.weather.gov/text.php?id=pacific.2011.03.11.073000

1.1 Observations

Ground instruments are normally placed in the vicinity of nuclear power plants to measure radioactivity and detect potential leaks. Additionally, the Comprehensive Test Ban Treaty Organization (CTBTO) maintains a network of worldwide stations that can detect radioactive clouds (United Nations General Assembly 1996; Monika 2012). When monitoring stations are located within a few kilometers downwind of the source, it is possible to estimate the radioactive release rate based on the measurements.

In the case of Fukushima, the instruments located in the vicinity of the power plant were ground gamma-ray counters and dust samplers, which are designed to detect small amount of radiation. However, the data are missing due to earthquake/tsunami damage and electrical power outages. In addition, there were site evacuations on March 15 (cited in Japanese government reports) that also led to gaps in the plant monitoring data (Ohba 2012).

Personnel and equipment from the US Department of Energy (DOE) National Nuclear Security Administration (NNSA), joined by additional members from DOE and Department of Defense (DOD), were quickly dispatched to Japan. The surveys used both fixed-wing aircraft as well as helicopters to collect measurements in the area surrounding the nuclear power plant. The first measurements were made on 18 March 2011, and therefore do not include the first large release which occurred on 15 March 2011 (Kreek 2012).

The Japanese power plant operators from the Tokyo Electric Power Company (TEPCO) pumped seawater directly into the reactor to regulate temperature and pressure. This delicate and difficult operation prevented a total core meltdown, but required the contaminated water to be discharged directly into the ocean (Matsunaga 2012). Radioactivity in water was also measured in the aftermath of the accident, showing higher levels than normal, but it is not possible to relate the water radioactivity to the atmospheric discharge.

In practice, it is not possible to estimate the Fukushima release rate using measurements alone.

1.2 Nuclear Reactor Modeling

Nuclear reactor models have been developed to simulate reactor failures, and compute the expected radiation dose. The Pacific Northwest National Laboratory (PNNL) and the US Nuclear Regulatory Committee jointly developed the Radiological Assessment System for Consequence Analysis (RASCAL) dose assessment system (Athey et al. 1993; McGuire et al. 2007). Given a specific type of reactor, containment vessel and amount of fuel available, the system simulates different scenarios that can lead to partial or full core melt. However, the model only simulates the amount of radiation inside the containment vessel. The radiation can potentially leak into the environment in several ways, such as by controlled venting to reduce pressure in the containment vessel, by small leaks in the structure, or by direct emission into the atmosphere in the case of a massive failure where the core is partially or fully exposed.

Ramsdell (2012) provides a general overview of estimated release amounts for Fukushima available and provides a comparison with a U.S. nuclear power plant surrogate using the RASCAL model. He concludes that about 2% of the core material in reactor 1, and about 1% in reactors 2 and 3 were estimated to have leaked into the environment. These figures are in good agreement with the estimates provided by TEPCO (Matsunaga 2012).

In general, it is possible to simulate the total amount of radiation leaked using a nuclear reactor model, but not its temporal release rate.

1.3 Atmospheric Modeling

The estimation of the release rate can be considered a special case of the general source detection problem. This class of methods use a combination of concentration measurements and numerical transport and dispersion simulations to reconstruct the source characteristics. The goal is to determine the source characteristics that minimize the error between simulated and measured concentrations. The assumption is that when the error is small, the characteristics of the source have been correctly identified. There are of course uncertainties associated with the numerical model used, the concentration measurements, the terrain characteristics, and the meteorological data.

Kathirgamanathan et al. (2004) reconstructed the continuous release rate over a period of time from a source at a known location using a least-squares minimization of the solution of an advection-diffusion equation.

Senocak et al. (2008) proposed a source detection methodology based on Bayesian inference and Markov chain Monte Carlo to estimate the turbulent diffusion parameters in a forward Gaussian plume dispersion model. The model accounts for zero and non-zero concentration measurements, and was validated with real and synthetic dispersion experiments. Haupt et al. (2007), Delle Monache et al. (2008), and Cervone and Franzese (2011) use forward numerical simulations from candidate sources, and employ different search strategies to identify the source that minimizes the error between simulated and observed concentrations.

Winiarek et al. (2011b) minimized the error between measurements and concentrations simulated using Eulerian dispersion models, based on the advection diffusion transport equation. A general methodology consisting of a sequential data assimilation algorithm was presented for the semi-automatic sequential reconstruction of a plume, and validated using ground concentrations measured in France and Finland. The same methodology was applied by Winiarek et al. (2011a) and Boquet (2012) to the reconstruction of the Fukushima accident, showing a good correlation between simulated and observed concentrations of the radioactive leaks of Caesium-134 and Caesium-137.

Schöppner et al. (2011) and Stohl et al. (2012) estimated the source of the Fukushima accident using global Comprehensive Nuclear-Test-Ban Treaty (CTBT) radionuclide data, assuming that the Fukushima accident was the only source of these radionuclides. A backward propagation model was used to determine the source-receptor sensitivity related to the adjoint concentration. The source was reconstructed daily for both Caesium-137 and Iodine-131 for the period ranging from March 10 to 30. The source release rate is identified with a daily temporal resolution, and there is good agreement between observations and simulations.

In the immediate aftermath of the accident, the National Atmospheric Release Advisory Center (NARAC) at LLNL performed operational dispersion simulations[2] and attempted to reconstruct the source term by performing spatio-temporal analysis between measurements and model values under different assumptions such as time-varying vs. constant release rates (Sugiyama and Nasstrom 2012). A comprehensive report of their simulations and results can be found in Sugiyama et al. (2012).

In this paper, a new method is proposed for the reconstruction of non-steady sources. The approach is based on forward numerical dispersion models and stochastic search to minimize the error between the observations and the simulations. This work extends to non-steady sources the authors' prior work on steady source estimation, where evolutionary algorithms were employed to drive a search process that identifies the characteristics of an unknown atmospheric release (Cervone and Franzese 2010, 2011; Cervone et al. 2010b,a).

The proposed method is applied to the reconstruction of the time-dependent source release rate of radiation leaked from the Fukushima reactors for a period of 5 days with a 2-h temporal resolution. This work was first presented at the Workshop: Methods for Estimating Radiation Release from Fukushima Daiichi, which took place at the National Center for Atmospheric Research (NCAR) in February 2012. The website for the event, which includes presentations of several of the works cited in this article can be found at: http://www.ral.ucar.edu/nsap/events/fukushima/.

2 Source Reconstruction Methodology

The proposed methodology identifies the source release rate which minimizes the error between observed and simulated concentrations.

In order to be able to reconstruct an unsteady release rate, a continuous release with a virtual constant rate q is discretized into a sequence of N consecutive finite-duration releases Q_n, with $n = 1 \ldots N$. Figure 1a shows a sample steady plume represented as a sequence of identical releases with the same rate q (the area of each release is constant). A time-varying release rate Q_n (Fig. 1b) can be obtained by multiplying q by a scalar w_n for each release n. The goal of the reconstructing procedure is to determine the vector $W = \{w_1 \ldots w_N\}$. The vector W

---

[2]https://str.llnl.gov/JanFeb12/sugiyama.html

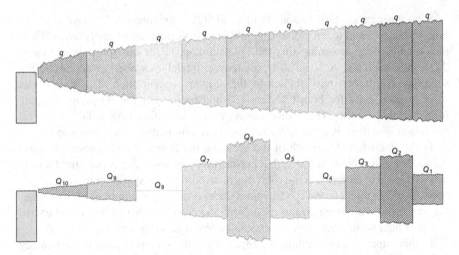

Fig. 1 Multiple consecutive releases of finite duration: releases of same rate q (*top*); discretized time-varying releases of different rate Q_n (*bottom*)

is identified through a stochastic optimization process that minimizes the error between the radioactivity levels measured at different locations in the domain, and the radioactivity simulated using a transport and dispersion model as described below.

2.1 Problem Definition

Time-dependent radioactivity (or concentration of radioactive material) is simulated at each ground location where radioactivity measurements are available. Specifically, at each location $x = (x, y, z)$ and time t the simulated total concentration of radioactive material C is equal to:

$$C(x,t) = \sum_{n=1}^{N} c_n(x,t) \tag{1}$$

where c_n is the simulated concentration at location x and time t generated by release n.

The goal is to find the unknown set of release rates Q_n which generates the field c according to (1). First, the space is discretized into M stations, and the sampling time t into K intervals. Then, a dispersion simulation is performed using a temporal sequence of N releases with equal mass rate q. Each one of the N releases generates a concentration ξ_{nmk}, where $n = 1 \ldots N$, $m = 1 \ldots M$, and $k = 1 \ldots K$.

Namely, ξ_{nmk} is the concentration generated by release n at location m at time k. The concentration generated by the real release rate Q_n can be written as

$$c_{nmk} = w_n \, \xi_{nmk} \tag{2}$$

where w_n are the N elements of an unknown vector W. The concentration corresponding to the real temporal sequence of releases Q_n can be written as:

$$C_{mk} = \sum_{n=1}^{N} c_{nmk} = \sum_{n=1}^{N} w_n \xi_{nmk} \tag{3}$$

Assuming a linear relationship between concentration and release rate, we can write

$$Q_n = w_n q \tag{4}$$

The unknown scalars w_n are calculated by minimizing the mean square difference between simulated concentration C_{mk} and observed concentration C_{mk}^o at location m and time k, over all the locations M and times K:

$$\Delta = \frac{1}{M + K} \sum_{m=1}^{M} \sum_{k=1}^{K} (C_{mk} - C_{mk}^o)^2 \tag{5}$$

In other words, the error Δ is the cost function that has to be minimized by optimizing the w parameters.

2.2 Minimization of the Error

Evolutionary algorithms (e.g. Bäck 1996) are particularly suited for this type of global spatio-temporal optimization, where vectors of correlated variables are optimized concurrently, and have already been applied successfully for atmospheric source characterization problems (Haupt et al. 2006; Cervone et al. 2010a). Strengths and weaknesses of the many types of evolutionary (or genetic) algorithms depend on the application. Our methodology is based on a class of evolutionary algorithms called Evolutionary Strategies, or ES (Rechenberg 1971; Schwefel 1974), which proved to be a good overall performer for atmospheric source detection problems (Cervone et al. 2010a).

While evolutionary algorithms in general are heuristics based on biologically in-spired iterative processes, ES address continuous parameter optimization problems in particular. Formally, if N is the number of optimized parameters w_n, and D_n are their domains, then the evolutionary algorithm attempts to minimize a goal function $\Delta(w_{1:N})$.

Central to the terminology of an evolutionary algorithm is the concept of *potential solution*, or *candidate solution*. Each solution contains a value for every one of the optimized parameters w_n. Like all evolutionary algorithms, ES maintain a *population* of potential solutions and attempt to improve on it interactively.

The process of randomly generating the initial population consists of assigning to each of the optimized parameters w_n random values uniformly sampled from D_n.

The particular type of ES used in this paper is known as $\text{ES}(\mu + \lambda)$ in which μ current solutions act as parents and are used to produce λ offspring that compete with their parents for survival. At each iterative step, only the best μ solutions (parents and offspring) are maintained, the rest being discarded.

Producing offspring from parents involves cloning, followed by the application of a perturbing (mutation) operator that induces minor stochastic variation to one or more variables. Therefore the search process can optimize multiple variables concurrently. When optimizing on a continuous domain, most of the times a Gaussian mutation is used. The degree of variation induced by this operator is tuned by its standard deviation, the meta parameter σ_n. One of the most delicate aspects is quantifying and controlling the magnitude of the stochastic variations. The global parameters σ_n are initialized to $1/6$ the size of D_n, and is adjusted throughout the optimization depending on the convergence rate (Fig. 2).

3   Models and Data

3.1   SCIPUFF T&D Model

The release reconstruction methodology can be used with any T&D model. The T&D model used in this study is the Second-order Closure Integrated Puff model (SCIPUFF), a Lagrangian puff dispersion model that uses a collection of Gaussian puffs to represent an arbitrary, three-dimensional, time-dependent concentration field (Sykes et al. 1984; Sykes and Gabruk 1997).

3.2   Radiation Data

The radiation data over the Japanese territory was collected via the System for Prediction of Environment Emergency Dose Information (SPEEDI),[3] which is maintained by the Nuclear Safety Division of the Japanese Ministry of Education, Culture, Sports, Science and Technology Disaster Prevention Network for Nuclear

---

[3]http://www.bousai.ne.jp/eng/index.html

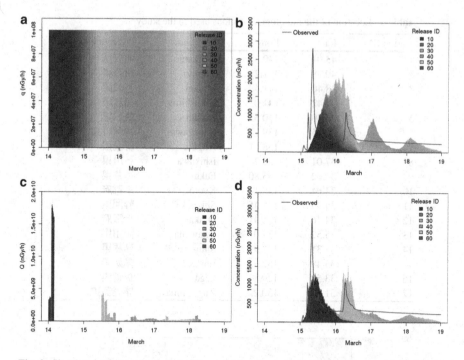

Fig. 2 Simulated release rate in Fukushima (*left*) and concentrations in Kanagawa (*right*). *Top panels* – results before optimization; *bottom panels* – results after optimization. The horizontal axis represents time from 14 to 19 March 2011. (**a**) Release rate in Fukushima before optimization. (**b**) Concentration in Kanagawa before optimization. *Solid line* – observations; *shaded area* – simulation. (**c**) Release rate in Fukushima after optimization. (**d**) Concentration in Kanagawa after optimization. *Solid line* – observations; *shaded area* – simulation

Environments. The network is used for real-time dose assessment in radiological emergencies, and it was instrumental in determining the risk areas evacuated by the Japanese Government at the time of the accident. The data consists of measurements of radiation updated at intervals of 10 min, and are collected at 218 stations grouped in 17 prefectures located across Japan.

Table 1 and Fig. 3 show the locations of the stations and their grouping into prefectures. The data for the prefectures of Fukushima (# 4) and Ibaraki (# 5), which are the ones closest to the release location, are not available. The data for each station was averaged by prefecture, and used in the experiments. The background radiation for each prefecture was removed by averaging data for 1 month prior to the accident, and removing the resulting value from the signal.

The stations are not uniformly distributed throughout Japan, but are concentrated in specific regions of each prefecture. Representative data for each prefecture are calculated by averaging the measurements of all the stations in the prefecture.

Table 1 Location of the radiological stations used in the study

Station ID	Lat.	Long.	Prefecture	Prefecture (Jap)
1	43.04	140.52	Hokkaido	北海道
2	40.98	141.27	Aomori	青森県
3	38.40	141.48	Miyagi	宮城県
4	37.42	141.03	Fukushima	福島県
5	36.40	140.53	Ibaraki	茨城県
6	35.35	139.71	Kanagawa	神奈川県
7	37.42	138.61	Niigata	新潟県
8	37.01	136.74	Ishikawa	石川県
9	35.62	135.80	Fukui	福井県
10	34.64	138.14	Kyoto	京都府
11	35.49	135.45	Shizuoka	静岡県
12	34.46	135.41	Osaka	大阪府
13	35.52	133.01	Okayama	岡山県
14	35.31	133.93	Shimane	島根県
15	33.48	132.32	Ehime	愛媛県
16	33.50	129.83	Saga	佐賀県
17	31.83	130.22	Kagoshima	鹿児島県

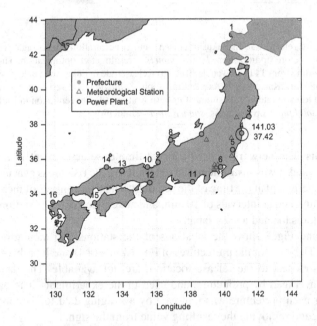

Fig. 3 Map of the simulation domain, showing the location of the radiation and meteorological measurements. The location of the Fukushima power plan is indicated with a *circle*. The horizontal axis shows degree longitudes East, and the vertical axis shows degree latitudes North

Table 2 Location of the ground meteorological stations used in the study

Station ID	Lat.	Long.	Elevation (m)	Airport
RJFS	37.20	140.40	375	Yes
IMIYAGIS3	38.39	140.72	541	No
INIIGATA1	37.11	138.92	145	No
RJAH	36.18	140.42	35	Yes
RJTT	35.55	139.78	8	Yes
IKANAGAW1	35.44	139.37	24	No
ITOKYOHI1	35.66	139.40	102	No
ITOKYOSE3	35.66	139.59	50	No
IU6771U42	35.52	139.48	71	No

3.3 Ground Meteorological Data

Ground meteorological measurements were obtained for nine locations, selected among those available around the Fukushima nuclear power plant, and in the prefectures for which radioactivity data are available. The data are available with a temporal resolution of 30 min. Table 2 shows the location information relative to the ground meteorological stations used in this study. The meteorological data used include observations for wind direction and speed, temperature, pressure, and rainfall information. All these measurements, including rainfall estimation, were converted and used in the SCIPUFF simulations.

3.4 Model Meteorological Data

The meteorological data used for the vertical profiles are from the NCEP Reanalysis II model. The model has a spatial resolution of 2.5° and a temporal resolution of four times daily (00:00Z, 06:00Z, 12:00Z, 18:00Z). Data for wind speed and direction, temperature and relative humidity at 17 pressure levels (hPa): 1000, 925, 850, 700, 600, 500, 400, 300, 250, 200, 150, 100, 70, 50, 30, 20, 10, were used by SCIPUFF.

3.5 Terrain Data

The Digital Elevation Model (DEM) terrain data used for the simulations is from the Global Land 1-km Base Elevation Project (GLOBE). The data has 1 km spatial resolution, and is derived from satellite observations and ground measurements. It is maintained by an international consortium of government agencies and universities, and distributed free of charge.

4   Results

The methodology described is applied to identify the source release rate associated with the Fukushima nuclear power plant accident. The time dependent release rate is determined for a period of 5 days, from 14 to 18 March 2011, with a 2-h resolution. As described in Sect. 2, a 5-day long continuous release of constant emission rate q was discretized into 60 releases, each of the duration of 2 h (Fig. 1, top). The SCIPUFF model was used to simulate all 60 releases.

Simulated and observed concentrations were compared at the locations identified in Table 1 and Fig. 3, and the evolutionary optimization algorithm is applied to identify the 60 unknown parameters w_n which minimize the error between the simulated and observed concentrations at each prefecture and each time step. This is a parallel optimization, in which the error is simultaneously minimized for all releases (Equation (5)). The time-dependent release rates Q_n is finally obtained according to Equation (4).

Note that the procedure requires only a single dispersion simulation of the initial 60 releases with constant q. Each release was uniquely associated to a specific fictitious material, so that at each location and time the total concentration can be computed as the additive contribution of the individual releases. Therefore the process is equivalent to simulating the release of 60 different gases.

The temporal resolution of the model output is 10 min, to match the temporal resolution of the observed concentration data. Although the temporal resolution is higher than the available resolution for the meteorological data, SCIPUFF interpolates between the available measurements and can reconstruct the field.

Figure 2a shows the initial constant q for the 60 releases over the 5 days considered in this study. The color indicates the release and ranges from shades of blue (releases 10–20) to green (releases 30–40) and red (releases 50–60).

Figure 2b shows the resulting concentration computed at the Kanagawa prefecture using the fixed release rate q. Each release is color coded using the same scheme of Fig. 2a, and it is thus possible to determine the individual contribution of each of the 60 releases as a function of time. Because the release rate is constant, the variation in the concentration is only due to the atmospheric transport and dispersion. The figure also shows the observed concentration as a continuous line. As expected, this non optimized case does not show a good agreement between measurements and simulation.

Figure 2c shows the reconstructed release rates Q_n as a function of time at Fukushima. Two major releases are identified for March 14th and 15th. The TEPCO time-series for the radiation dose measured at the main gate of the Fukushima nuclear power plant, and reported by Shozugawa et al. (2012), agrees with our findings.

Figure 2d shows the resulting simulated concentration at Kanagawa, obtained using the identified Q_n. The time-location of both peaks is accurately simulated. The simulation underestimates the magnitude of the first episode, and overestimates its time duration. Both magnitude and width of the second episode are more accurately captured.

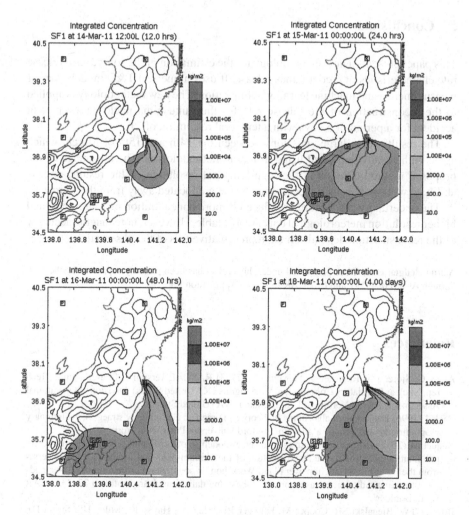

Fig. 4 Contour levels of the vertically integrated concentration simulated by the SCIPUFF model using the identified release rate. The panels show results at 12, 24, 48 and 96 h after March 14, 2011, 00:00 JST

Figure 4 shows the SCIPUFF simulation of the Fukushima accident using the reconstructed time-variable release rates Q_n. The figure displays contour lines of the vertically integrated concentration in kg m^{-2}. The terrain elevation is also indicated. The locations where meteorological vertical profiles and surface observations are available are indicated by the letters $\boxed{\text{P}}$ and $\boxed{\text{S}}$.

5   Conclusions

This paper presents a new methodology for the estimation of the source term release rate characteristics of a continuous release. It uses numerical T&D models, sensor measurements, and machine learning optimization. The new methodology is applied to the reconstruction of the release rate for the Fukushima nuclear power plant accident for a period of 5 days with a temporal resolution of 2 h.

The results show large emissions between the 14th and 15th of March, which agree with results published using different methodologies. The advantage of the proposed method consists in the high temporal resolution of the results, and its ability to work with sensor measurements that are located away from the source.

The uncertainty can be further reduced if more concentration measurements and higher resolution meteorological data are available. However, note that the accuracy of the reconstruction depends on the representativeness of the observations.

Acknowledgements Work performed under this project has been partially supported by the NSF through Award 0849191 and by NCAR Research Application Lab.

References

Athey G, Sjoreen A, Ramsdell J, McKenna T, (1993) RASCAL version 2. 0: user's guide. Tech rep, Nuclear Regulatory Commission, Washington, DC (United States). Division of Operational Assessment; Pacific Northwest Lab, Richland, WA (United States)

Bäck T (1996) Evolutionary algorithms in theory and practice: evolutionary straegies, evolutionary programming, and genetic algorithms. Oxford University Press, New York

Boquet M (2012) Estimation of errors in inverse modeling of accidental release of atmospheric pollutant: application to the reconstruction of the Cesium- 137 and Iodine-131 source terms from the Fukushima Daiichi power plant. Workshop on source term estimation (STE) methods for estimating radiation release from the fukushima daiichi nuclear plant, Feb 22–23, NCAR RAL, Boulder

Bowyer TW, Biegalski SR, Cooper M, Eslinger PW, Haas D, Hayes JC, Miley HS, Strom DJ, Woods V (2011) Elevated radioxenon detected remotely following the fukushima nuclear accident. J Environ Radioact 102(7):681–687

Calabrese E (2011) Improving the scientific foundations for estimating health risks from the Fukushima incident. Proc Natl Acad Sci 108(49):19447–19448

Cervone G, Franzese P (2010) Monte Carlo source detection of atmospheric emissions and error functions analysis. Comput Geosci 36(7):902–909

Cervone G, Franzese P (2011) Non-darwinian evolution for the source detection of atmospheric releases. Atmos Environ 45(26):4497–4506

Cervone G, Franzese P, Grajdeanu A, (2010a) Characterization of atmospheric contaminant sources using adaptive evolutionary algorithms. Atmos Environ 44:3787–3796

Cervone G, Franzese P, Keesee AP (2010b) Algorithm quasi-optimal (AQ) learning. WIREs Comput Stat 2(2):218–236

Delle Monache L, Lundquist J, Kosović B, Johannesson G, Dyer K, Aines R, Chow F, Belles R, Hanley W, Larsen S, Loosmore G, Nitao J, Sugiyama G, Vogt P (2008) Bayesian inference and Markov Chain Monte Carlo sampling to reconstruct a contaminant source on a continental scale. J Appl Meteorol Climatol 47:2600–2613

Haupt SE, Young GS, Allen CT (2006) Validation of a receptor/dispersion model coupled with a genetic algorithm using synthetic data. J Appl Meteorol 45:476–490

Haupt SE, Young GS, Allen CT (2007) A genetic algorithm method to assimilate sensor data for a toxic contaminant release. J Comput 2(6):85–93

Kathirgamanathan P, McKibbin R, McLachlan R (2004) Source release-rate estimation of atmospheric pollution from a non-steady point source at a known location. Environ Model Assess 9(1):33–42

Kinoshita N, Sueki K, Sasa K, Kitagawa J, Ikarashi S, Nishimura T, Wong Y, Satou Y, Handa K, Takahashi T et al (2011) Assessment of individual radionuclide distributions from the Fukushima nuclear accident covering central-east Japan. Proc Natl Acad Sci 108(49): 19526–19529

Kreek S (2012) Description of observations collected around the Fukushima site by the US DOE teams. Workshop on source term estimation (STE) methods for estimating radiation release from the Fukushima Daiichi nuclear plant, Feb 22–23, NCAR RAL, Boulder

Masson O, Baeza A, Bieringer J, Brudecki K, Bucci S, Cappai M, Carvalho F, Connan O, Cosma C, Dalheimer A et al (2011) Tracking of airborne radionuclides from the damaged Fukushima Daiichi nuclear reactors by European networks. Environ Sci Technol 45(18):7670–7677

Matsunaga T (2012) Introduction of Fukushima Daiichi nuclear power station accident. Workshop on source term estimation (STE) methods for estimating radiation release from the Fukushima Daiichi nuclear plant, Feb 22–23, NCAR RAL, Boulder

McGuire S, Ramsdell J, Athey G (2007) RASCAL 3.0. 5: description of models and methods. Tech rep, Office of Nuclear Security and Incident Response, US Nuclear Regulatory Commission

Monika K (2012) CTBTO radionuclide detections in the aftermath of the Fukushima release and a necessity for improved source inversion algorithms. Workshop on source term estimation (STE) methods for estimating radiation release from the Fukushima Daiichi nuclear plant, Feb 22–23, NCAR RAL, Boulder

Ohba R (2012) Report on a recent field program to collect radiation measurements surrounding the Fukushima nuclear power station. Workshop on source term estimation (STE) methods for estimating radiation release from the Fukushima Daiichi nuclear plant, Feb 22–23, NCAR RAL, Boulder

Potiriadis C, Kolovou M, Clouvas A, Xanthos S (2012) Environmental radioactivity measurements in Greece following the Fukushima Daichi nuclear accident. Radiat Prot Dosimetry 150(4):441–447

Ramsdell V (2012) Modeling the physical processes that impact the fate and fall-out of radioactive materials. Workshop on source term estimation (STE) methods for estimating radiation release from the Fukushima Daiichi nuclear plant, Feb 22–23, NCAR RAL, Boulder

Rechenberg I (1971) Evolutionstrategie – Optimierung technischer Systeme nach Prinzipien der biologischen Evolution. Ph.D. thesis, Technical University of Berlin

Schöppner M, Plastino W, Povinec PP, Wotawa G, Bella F, Budano A, Vincenzi MD, Ruggieri F (2011) Estimation of the time-dependent radioactive source-term from the Fukushima nuclear power plant accident using atmospheric transport modelling. J Environ Radioact. http://dx.doi.org/10.1016/j.bbr.2011.03.031

Schwefel H-P (1974) Numerische optimierung von computer-modellen. Ph.D. thesis, Technical University of Berlin

Senocak I, Hengartner N, Short M, Daniel W (2008) Stochastic event reconstruction of atmospheric contaminant dispersion using Bayesian inference. Atmos Environ 42(33):7718–7727

Shozugawa K, Nogawa N, Matsuo M (2012) Deposition of fission and activation products after the fukushima dai-ichi nuclear power plant accident. Environ Pollut 163:243–247

Stohl A, Seibert P, Wotawa G, Arnold D, Burkhart J, Eckhardt S, Tapia C, Vargas A, Yasunari T (2012) Xenon-133 and caesium-137 releases into the atmosphere from the fukushima dai-ichi nuclear power plant: determination of the source term, atmospheric dispersion, and deposition. Atmos Chem Phys 12(5):2313–2343

Sugiyama G, Nasstrom J (2012) NARAC source reconstruction during the response to the Fukushima Daiichi nuclear power plant emergency. Workshop on source term estimation (STE) methods for estimating radiation release from the Fukushima Daiichi nuclear plant, Feb 22–23, NCAR RAL, Boulder

Sugiyama G, Nasstrom J, Pobanz B, Foster K, Simpson M, Vogt P, Aluzzi F, Homann S (2012) Atmospheric dispersion modeling: challenges of the fukushima daiichi response. Health Phys 102(5):493–508

Sykes RI, Gabruk RS (1997) A second-order closure model for the effect of averaging time on turbulent plume dispersion. J Appl Meteorol 36:165–184

Sykes RI, Lewellen WS, Parker SF (1984) A turbulent transport model for concentration fluctuation and fluxes. J Fluid Mech 139:193–218

United Nations General Assembly (1996) General assembly resolution number 50/245 on 10 september 1996. Tech rep, United Nations

Winiarek V, Bocquet M, Roustan Y, Birman C, Tran P, (2011a) Atmospheric dispersion of radionuclides from the Fukushima-Daichii nuclear power plant. CEREA, joint laboratory École des Ponts ParisTech and EdF R&D

Winiarek V, Vira J, Bocquet M, Sofiev M, Saunier O (2011b) Towards the operational estimation of a radiological plume using data assimilation after a radiological accidental atmospheric release. Atmos Environ 45(17):2944–2955

Yasunari T, Stohl A, Hayano R, Burkhart J, Eckhardt S, Yasunari T (2011) Cesium-137 deposition and contamination of Japanese soils due to the Fukushima nuclear accident. Proc Natl Acad Sci 108(49):19530–19534

GIS-Based Traffic Simulation Using OSM

Jörg Dallmeyer, Andreas D. Lattner, and Ingo J. Timm

Abstract This chapter demonstrates how to build up a traffic simulation on the base of a Geographic Information System (GIS). Maps from the OpenStreetMap (OSM) initiative have shown to be appropriate for usage in this field. Essential steps from OSM over GIS to a graph data structure for use in traffic simulation are described. The work is done with the focus on urban scenarios. The crucial decision, which types of road users to integrate into a simulation and how to model them, is discussed. A case scenario shows the utility of data mining techniques in the field of traffic simulation. The scenario aims at predicting traffic jams in the city of Frankfurt am Main with help of a learned classifier. Our results show that taking into account simple and partial information about the traffic situation can lead to a huge gain of knowledge when using data mining techniques in the face of predicting of traffic situations.

Keywords Traffic simulation • Machine learning • Jam forecasting • Urban traffic

1 Introduction

Simulations are commonly used for understanding and prediction of traffic phenomena, traffic densities and traffic flows. It is much cheaper to test a new scenario in simulation before implementing it in reality. For a realistic simulation of urban

J. Dallmeyer (✉) • A.D. Lattner
Information Systems and Simulation, Institute of Computer Science, Goethe University
Frankfurt, P.O. Box 11 19 32, 60054 Frankfurt, Germany
e-mail: dallmeyer@cs.uni-frankfurt.de; lattner@cs.uni-frankfurt.de

I.J. Timm
Business Informatics I, University of Trier, D-54286 Trier, Germany
e-mail: ingo.timm@uni-trier.de

G. Cervone et al. (eds.), *Data Mining for Geoinformatics: Methods and Applications*,
DOI 10.1007/978-1-4614-7669-6_4, © Springer Science+Business Media New York 2014

scenarios, two elementary problems need to be solved. At first, the road structure needs to be modeled in a simulation graph. Second, the behavior of road users and traffic rules need to be modeled. According to the focus of the simulation, additional information needs to be integrated into the model (e.g., a digital terrain model or a model for gas consumption and CO_2 emissions).

Today, for almost any major city, very detailed road models exist in layers for Geographic Information Systems (GIS). GIS are very powerful tools in order to deal with heterogeneous data sources particularly with regard to building road models. With help of the renderer of a GIS, it is easy to show the effects in simulation directly. Therefore, a couple of GIS layers, specifying different types of geometry (e.g., roads and polygons), are rendered on top of each other and a typical map is produced.

The simulation of multimodal traffic is a challenge for traffic simulation systems. Interdependencies from, e.g., bicyclists with cars or cars with pedestrians need to be modeled. Most traffic simulations therefore either simulate traffic in a very simplified way and cannot be adapted realistically for urban scenarios (e.g., Transims (Nagel et al. 1999), MATSim (Rieser 2010)) or are high fidelity simulations which are not sufficient for large scenarios (e.g., VISSIM (Fellendorf 1994)). Simulations differ widely in the needed time effort for setting up a simulation.

After having built a traffic simulation system, traffic engineers need to evaluate scenarios with help of the system. Different output values can be used in order to compare different scenarios (e.g., traffic flows, mean velocities, travel times, time spent in front of red traffic lights, number of accidents, ...). A manual analysis of such data is often not feasible because of the high amount of information and numerous attributes. Methods from machine learning can be used to find interesting rules like "When the traffic density in the area x is bigger than v, traffic will get stuck in area y". Apparently, such rules are not trivially to be generated, but steps into this direction have been taken (e.g., Lattner et al. 2011).

The *OpenStreetMap* (OSM) project has grown constantly in the last years and is today a useful source for road maps for different purposes. It would be a benefit for traffic research to build traffic simulation models upon OSM. OSM maps are not directly usable for GIS. Thus, a transformation step has to be established.

This chapter gives an introduction on how to use OSM data for traffic simulation (Sect. 2) and how to build up a simulation graph from GIS information (Sect. 3). It also introduces the developed traffic models (Sect. 4) and a short introduction to machine learning for traffic simulation is provided (Sect. 5). A case scenario is discussed (Sect. 6) and the chapter ends with a short summary of the content and perspectives for future investigations (Sect. 7). The discussed architecture and models are part of the traffic simulation system MAINS[2]IM (MultimodAl INnercity SIMulation).[1]

---

[1]http://www.mainsim.eu

2   OpenStreetMap for Traffic Simulation

This section discusses OpenStreetMap[2] (OSM) and how to obtain data from the project. The design of cartographical material from OSM is described and an idea about how to extract important information from OSM is given.

OSM is a free of charge project giving access to cartographical material from all over the world. The maps are created by voluntary cartographers in the manner of Wikipedia. The data is licensed currently under Creative Commons Attribution-Share Alike 2.0 (CC-BY-SA),[3] but this could change in the future (Bennett 2010).

OSM has grown in the last years and has reached a level of detail and quality enabling it to be used for different purposes even though the data quality differs across different areas (Haklay 2010; Zielstra and Zipf 2010). For example, OSM maps are used in different smartphones for navigation purposes. In the area of Transportation Simulation, OSM is used as a data source in order to built up simulation graphs (Dallmeyer et al. 2011; Zilske et al. 2011).

OSM uses an XML-format. At first, an amount of nodes is defined. LineString and polygon geometries are built with help of way elements. These can be, e.g., roads, rivers, forests and local areas. ways store references to the nodes they consist of. The strengths of the format are its simplicity, memory efficiency, and extensibility. Additional information can be stored via usage of tag elements, which use a key and a value. In addition to nodes and ways, relations can be created, which store, e.g., bus routes or bicycle tracks. relations are always the last section of an OSM file. Figure 1 shows the basic elements.

OSM files can be exported directly from the OSM website. Only small areas can be exported. Another possibility is to download a map from a whole federal

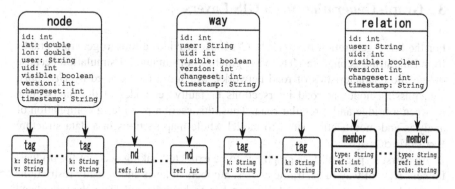

Fig. 1 Basic OSM elements

---

[2]http://www.openstreetmap.org

[3]http://www.openstreetmap.org/copyright, last visited December 15th, 2011

state from an external provider.[4] The files can easily have a size of more than 1 GB. Thus, we developed a way to clip the map by oneself to a polygon P. The strict order nodes $\rightarrow$ ways $\rightarrow$ relations can be used to do the clipping efficiently. The XML elements are parsed via a SAX parser and each node which is covered by P is stored in the clipped new output file. The remainder is dropped. Afterwards, each way, which contains only references to nodes, which are in the output file are attached to the clipped file, the rest is dropped. The same way, relations can be filtered.

In order to keep each way which is covered by P or crosses P, the OSM file has to be processed a second time to amend the nodes that have been dropped during the first parsing of the file though being used in one of the way elements, that are not fully covered by P.

For the purpose of traffic simulation, it is much easier to work with a layer based GIS than with a whole XML tree. This gives the advantages of a powerful rendering method for visualization of the simulation and to divide information which is used for simulation from information which is only used for rendering. Layers are a logical way of grouping associated information. The clipped OSM document is converted to a number of Shapefile layers.[5] This is done with help of the open source GIS toolkit *GeoTools*.[6]

GeoTools is written in Java. It provides classes for the handling of geoinformation and can be extended arbitrarily. In order to build up a traffic simulation system, it is necessary to extract a simulation graph data structure from the given geographical map. The interplay of different layers and specific characteristic for traffic simulation are discussed in the next section.

3   Graph Generation from GIS Layers

For the basic generation of a graph, *GeoTools* provides a tool to generate graphs from LineString geometries. The basic graph is not suitable for simulation because of specific characteristics of road networks, which are not covered by *GeoTools*, e.g., missing nodes at road intersections or faulty set nodes at the intersection between a bridge and a regular road. Thus, this section provides a compilation of analysis and modification steps to model whole map sections in a data structure ExtendedGraph.

As a first step, a faultless graph has to be extracted, taking account for bridges, tunnels, right of way rules, traffic circles, numbers of lanes and velocity restrictions (Sect. 3.1). Then, additional information has to be extracted from the calculated

---

[4]http://www.geofabrik.de/en/data/download.html

[5]Shapefile is a standard format for geographical information.

[6]http://www.geotools.org/

graph, assigning which road may be used by which type of road user, finding bus routes, calculating ways in the graph and assigning edge usage probabilities (Sect. 3.2).

3.1 Generating an ExtendedGraph

At first, all basic edges and nodes of the graph are converted into EdgeInformations (EI) and NodeInformations (NI), which use as much information from OSM as possible (e.g., number of lanes, type of road, velocity restrictions). Unavailable information is amended via lookup tables, with respect to the type of road. Typical values for a residential road would be #lanes $= 1$ and $v_{max} = 30$ km $\cdot$ h^{-1}. The connections between EIs are set: each EI is connected to two NIs NI$_a$ and NI$_b$ and each NI has at least the connection to one EI.

EIs are grouped in a structure called EdgeInformationCollection (EIC), NIs in a NodeInformationCollection (NIC). The collections give methods for selecting specific elements and manipulating EIs and NIs for different purposes.

At the current stage, each road is modeled with one geometry. Roads do not cross each other when they proceed after the crossing point. Therefore, intersections between EIs are detected and the corresponding EIs are split up at the intersection point. A new NI connects the crossing parts.

We are using a graph structure which stores one EI per road segment. The stored nodes of an OSM way are sorted in the direction of the course of the road. If a road is oneway, this attribute will be stored in the corresponding EI and the EI can only be used from NI$_a$ to NI$_b$ and not in the opposite direction. This means that the position of a simulated road user ru on the road depends on the NI, ru has visited last. In OSM, small traffic circles or turning areas are often stored with only one edge. In this case, EIs exist which have NI$_a$ = NI$_b$ and it is impossible for ru to decide in which direction it drives on the corresponding road. Thus, such EIs are split in the center of their geometries and new NIs are inserted.

NIs being connected to EIs representing bridges or tunnels are checked, whether they do connect EIs which are bridges or tunnels with EIs which are not. In this case, a new NI is inserted at the same position of the existing NI and the EIs which are bridges or tunnels are connected to the new NI and are removed from the old one. In our representation, crosswalks and traffic lights are stored in NIs and therefore, EIs are split whenever a traffic control has to be inserted. The corresponding positions are read from a GIS layer representing points.

Because v_{max} of each EI is set in relation to the type of road when no information is given, highways crossing towns result in higher values of v_{max} than in reality. From the layer of polygons, each inner-city area ica is checked for EIs being located in ica. The value of the maximum velocity is adjusted to the minimum of v_{max} and 50 km $\cdot$ h^{-1}, which is a typical value for the standard inner-city velocity. EIs being located in parking lots are detected identically.

In urban scenarios, traffic circles have gained an important role. The problem at this point is, that in OSM not all traffic circles are marked as such and therefore an analysis step is necessary to detect traffic circles. All parts of the traffic circle have identical values in respect to name, road type, maximum velocity, number of lanes and traffic circles are always oneway streets. The geometries of the parts of a traffic circle form a circle in anticlockwise direction. With the determined characteristics, traffic circles can be found in the graph.

OSM does not store information about which type of road user may use a road. A further lookup table is used to define this information in relation to different road types. In urban scenarios, the right of way is regulated by the priority of the road and the rule "right before left".[7] Each NI stores information about all connections over this NI. A connection has a direction (left, right, ahead) and an angle. The priority of a road is stored in the corresponding EI with help of an additional lookup table for the type of the EI. The higher the priority value, the higher right of way of EI. The highest priority is given to traffic circles without respect to their road type.

Up to this point, a consistent ExtendedGraph is built. The following subsection discusses additional information, which needs to be calculated, according to the subject of simulation.

3.2 Determining Additional Information

Bus routes can be stored in OSM via usage of relations. The problem at this point is, that routes are usually not stored from a starting point to an end point under usage of way objects in the sequential arrangement of their appearance on the route. But at least, the used ways can be read out and a heuristic algorithm which rearranges the parts of the route to a plausible aggregate route can be applied. Buses can be used in the simulation to let groups of pedestrians appear at bus stops and also for simulating public transportation where buses stop at bus stops. This leads to abruptly changing traffic situations in areas surrounding bus stops, because of pedestrians using a crosswalk, pushing the button of a traffic light or crossing the road at an insufficient gap.

Assuming that the simulation graph will not alter significantly during the simulation, all possible routes in the simulation graph are calculated. From each $NI_{start} \in NIC$, Dijkstra's algorithm (e.g., Cormen et al. 2001) is performed. The ID of the first NI_1 on the way from NI_{start} to each $NI_{dest} \in NIC \setminus NI_{start}$ is stored. After performing this algorithm, each NI_{start} holds an array, where the n-th entry is the ID of the first NI on the way to the NI_{dest} with ID n.

---

[7]The described method implements rules for right hand traffic. For left hand traffic, the right of way priorities need to be switched.

The method is performed for the three basic routing characteristics "car", "bicycle" and "pedestrian". Space reduction is achieved due to creating lookup tables for each NI, giving the stored IDs shorter values. This is possible due to the fact, that each NI has only as much different IDs to store, as it has connected EIs.

In order to have random behavior in routing, the calculated routes are passed to an analysis step. All routes are run from NI_{start} to NI_{dest} and each NI counts, how often its connected EIs is used on the runs. Edge usage probabilities are set up with respect to these counts. Three different routing mechanisms can now be used during simulation: Precalculated routes, online calculated routes (which need more computational time, but can deal with changes in the ExtendedGraph) and random walks with respect to precalculated edge usage probabilities (Dallmeyer et al. 2011).

The computed ExtendedGraph can be stored into a file and directly used as input for simulation runs.

After having calculated the ExtendedGraph, it is necessary to decide, which kind of road users to simulate and how to model, e.g., cars and pedestrians. The next section addresses this crucial factor.

4 Simulation Models

The focus of this work is the simulation of multimodal traffic in urban scenarios. Each group of road users needs different behavioral models. This section gives a brief introduction of how our models work. It is important to model cars (e.g., passenger cars, trucks and buses; Sect. 4.1), bicycles (Sect. 4.2) and pedestrians (Sect. 4.3).

4.1 Space Continuous Car Model

The simulation of car traffic is often done with help of the Nagel-Schreckenberg model (NSM) (Nagel and Schreckenberg 1992), which is a cellular automaton model for freeway traffic. It is the de facto standard, because of its simplicity and, nevertheless, capability to model freeway traffic in a realistic way with respect to macroscopic properties. NSM divides a road into cells with length 7.5 m. Cars may have discrete velocities in the range $v \in \{0 \cdots 5\} \left[cell \cdot s^{-1} \right]$. The model is discrete in time. One simulation iteration corresponds to 1 s real time.

Each simulated car performs the following steps in parallel. At first, try to accelerate ($v \leftarrow v + 1$), if $v < 5$. Determine the gap γ to the preceding car. Set $v \leftarrow \min(v, \gamma)$. Dally with probability p. When dallying, set $v \leftarrow \max(0, v - 1)$.

It was shown that emerging phenomena through interdependencies of many cars like *phantom jams*[8] can be simulated with NSM. Furthermore, the model could be calibrated to measurement data from real freeways. Many work has been done in the community of traffic simulators, e.g., extending the model with brake lights (Hafstein et al. 2003), a slow start rule (Helbing 1997), smaller cell sizes (Krauss et al. 1996) and multi lane traffic (Knospe et al. 2002).

Although NSM provides many extensions, it has been reasonable to remove the cells from the roads and to build a space continuous car model which uses the basic functionalities of NSM, but being able to build arbitrary acceleration and deceleration functions. One main reason for space continuity are different velocity spreadings for different types of road users. For example, a bicycle usually moves much slower than a car. Any length of car can be modeled. Interaction on the roads can be simulated in more detail. This could only be done in NSM when using very small cells, but this would decrease the computational advantages of NSM significantly.

The model uses the same update steps as NSM but with variable velocity functions. Different modifications and enhancements, described in literature are built in our model, too. We could show, that the basic model reproduces macroscopic data on freeways (Dallmeyer et al. 2011). Cars are able to change lanes and to choose lanes for turning maneuvers.

4.2 Bicycle Model

Bicycles are currently not under detailed investigation in science. But there are publications about how bicycles behave in urban traffic (e.g., Johnson et al. 2011; Larsen and El-Geneidy 2011). The investigated bicycle model is built on determined velocity distributions. Bicycles have mainly the same behavioral model as cars, but with different dallying properties and, of course, different routes. Bicycles are interesting for traffic simulation because of the interaction with cars.

Cars may overtake bicycles when it is safe. Each car determines, what is seen to be *safe*. Overtaking is done with respect to the desire to overtake, the gap to oncoming traffic, the possibility to get back to the right lane after overtaking (there needs to be a gap, too), the distance to the next road junction and the width of the road (with respect to the width of the car, the oncoming car and the bicycle).

Another occurrence in urban traffic is the pushing to the front of cyclists to red traffic lights, overtaking on the left, on the right or both. This seems to be intended by traffic planners in some cases (e.g., if there is a specific region for cyclists in front of the stop line for cars). When the light switches to green, cars need to overtake the bicycles again. Not all bicycles overtake at red lights. This can be modeled with help of a probability function.

---

[8] A phantom jam is a jam which occurs without an obvious reason. It is build up from dallying and overdosed braking in NSM and has likewise reasons, in reality (Helbing 2001).

4.3 Pedestrian Model

The simulation of pedestrians has been given little attention in the past (Ishaque and Noland 2008), even though every human is a pedestrian. Pedestrians walk on sidewalks and cross roads, when it is necessary. Traffic lights and crosswalks are preferred for crossing the road (crossing at NIs) compared to crossing without right of way (crossing at EIs). It is assumed that pedestrian velocity will average over time and thus, that the mutual influence of pedestrians while walking on sidewalks (EIs) can be neglected. It is more interesting to determine, how long the crossing of a road will take. A model accounting for interaction between pedestrians is used for crossing roads at NIs.

The cellular automaton model for pedestrian movement, presented in Blue and Adler (2001) is used as basis for a space continuous model enabling pedestrians to avoid other pedestrians at NIs, walk with velocities in relation to the free space on NIs and to choose lanes. For example, a crosswalk has a width w_c and each pedestrian has a width w_p. The crossing then has $\left\lfloor w_c \cdot w_p^{-1} \right\rfloor$ lanes. The underlying principles are used for individual velocities with respect to different types of pedestrians (e.g., children, grown-ups and seniors). The velocities of pedestrians vary over different situations (e.g., walking on the sidewalk, crossing a road without right of way or crossing the road on a crosswalk).

Each simulated pedestrian accepts individual gaps in road traffic for crossing a road. This is done after the concept of Estimated Crossing Time (ECT) (Ottomanelli et al. 2009), which takes into account of individual aggressiveness of pedestrians. It may happen that a pedestrian misjudges a gap and is not able to finish crossing a road before a car or bicycle arrives at its position. Pedestrians are visible for road traffic only when crossing a road and then are treated like normal standing cars. An approaching car will slow down and in extreme cases come to standstill until a crossing pedestrian has left the road.

On the other hand, pedestrians cross NIs. They can do so without having the right of way. They decide whether crossing is appropriate with respect to traffic. Whenever a pedestrian is crossing an NI at one side, all road users planning to cross this side, will wait until the crossing is done. Pedestrians pass crosswalks without respect to traffic. Pedestrians are able to virtually push the buttons of traffic lights, forcing them to turn red for road traffic during the next simulated minute. The mentioned interactions influence urban traffic. A detailed description of the pedestrian model will be available in a separate publication (Dallmeyer et al. 2012a).

Different parameters can be measured during simulation runs, e.g., average velocities, traffic densities or traffic flows. In order to spot relationships between different parameters, appropriate techniques have to be applied to the system. The next section gives a brief introduction on machine learning and how to use it for traffic simulation.

5 Learning in Traffic Scenarios

Different learning paradigms can be applied to simulated traffic scenarios in order to identify regularities or useful behaviors. In this section, we discuss different options following the classification of learning approaches regarding learning feedback, namely supervised learning, unsupervised learning, and reinforcement learning (see, e.g., Russell and Norvig 2003).

The evaluation scenario in the subsequent section addresses the question how to predict if a congestion occurs at a certain road segment using information about the number of road users at different regions in the city. We also refer to this example in order to illustrate the different learning approaches.

In supervised learning, the desired output of the concept to be learned is known in advance. In the case of supervised learning from examples, for instance, each example in the training data consists of a set of features as well as a corresponding class. Referring the aforementioned scenario, features can be, e.g., the number of cars in different residential areas and the target class, if this situation has lead to a congestion at a specific road segment. The learning task is to generate a hypothesis which – hopefully – represents well the true underlying concept of the data. Learning results in a classifier which can be used to classify (previously unseen) examples. Having captured the number of cars in the different residential areas, the classifier can predict if a congestion is expected at the point of interest. It is aimed at generating a general classifier which does not only cover the training data well but additionally performs well on unseen examples. Approaches to supervised learning are, among others, decision tree learning, decision rule learning, learning in neural networks, and Bayesian learning. We focus here on symbolic learning approaches (decision tree learning and decision rule learning) as the generated classifiers lead to a comprehensible representation. Figure 2 illustrates the process of supervised learning from traffic simulation experiments.

In contrast, in unsupervised learning no information about the desired output is provided. The learning task is to identify regularities in the data (without target concept) or to group examples with respect to their similarity. In the first case, (sequential) association rules can be identified (e.g., Agrawal and Srikant 1994; Pei et al. 2001): If there is a high traffic density in area 1 and area 2, there will also be a high traffic density in area 3. In the latter case, the data would be grouped in a way that a classification scheme is generated, for instance, grouping similar traffic situations using a specified similarity measurement. This process is called clustering (see, e.g., Jain et al. 1999). Getting an overview of different groups in the data can help to analyze and to come up with certain strategies for the identified groups.

No direct feedback is provided to the learner in reinforcement learning. The general setting in reinforcement learning is, that an agent can perceive the current situation and can decide what action to take. In dependence of the actions, a reward will be provided to the agent. However, this reward is not necessarily instantaneous and might also depend on further activities or events beyond control of the agent. Referring to the traffic scenario once again, a reinforcement learning setting could

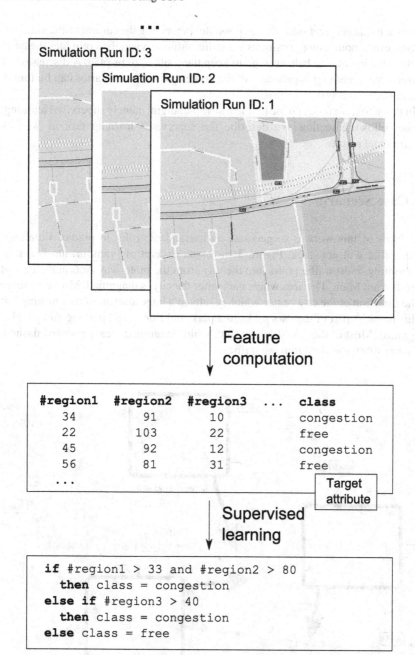

Fig. 2 Supervised learning in traffic simulation

be to reach a target position as fast as possible. Perceiving the current traffic situation (maybe even about remote positions via radio traffic service) then provides the basis for the decision how to behave, e.g., to keep the route or to re-plan. A discussion of reinforcement learning approaches in the context of traffic scenarios can be found, e.g., in Bazzan (2009).

In this work, we focus on the first learning paradigm, namely supervised learning. In the following section, we describe the supervised learning task as well as experimental results.

6 Case Scenario

The focus of this work is to generate understandable rules to predict situations, when traffic will get stuck. Figure 3 shows an excerpt of Frankfurt am Main. In the morning, most traffic pours into the city from the motorway A66 in the east of Frankfurt am Main. The area where cars enter the city is magnified. Most cars drive to the direction of the city center which results in a huge amount of cars turning left at this point. A part of the cars parks in a park-and-ride (P+R) parking garage, also magnified. Most of the cars drive on to the third magnified area, presented dashed. This area often has slow-moving traffic.

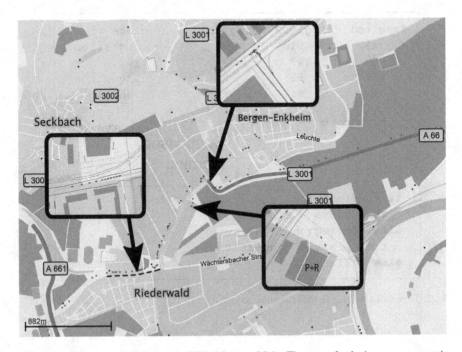

Fig. 3 Case Scenario: OSM excerpt of Frankfurt am Main. The area of velocity measurement is shown *dashed*

This results from a huge amount of unsynchronized traffic lights and the exceeded road capacity limit in this area. The traffic flow is influenced by pedestrians pushing the buttons of traffic lights. It is assumed to be possible to count the number of cars in a given area. This could be done with help of induction loops at central crossing points and extrapolation of these counts for a whole area. It is more difficult to count the number of bicycles, because of the possibility to additionally use cycle ways and paths. Pedestrians cannot be counted easily. It would be possible to count how often pedestrians push the buttons of traffic lights during a given time interval in a defined area.

In the case scenario, only cars are counted. The urban districts "Bergen-Enkheim", "Seckbach" and "Riederwald" are observed independently. In addition, the motorway A66 in the direction of the city of Frankfurt am Main gets observed. After a warm-up phase of the simulation (2,000 iterations), the number of cars in the measurement areas are counted ($\#Riederwald, \#Bergen-Enkheim, \#A66, \#Seckbach$) and the sum of them is calculated. These values are used as snapshots. In traffic scenarios, *time* is always an important parameter. The situation at a time t might influence the situation at time $t + \Delta$ in another region. The value of Δ is chosen $\Delta = 300\,$s, which is 5 min real-time. The value is chosen arbitrarily and is used as a first try, estimating the travel time from the areas of measurement to the point of interest. After Δ, the mean velocity $\bar{v}$ of all cars in the dashed jamming area is measured over a time of 60 s (1 min). Note that the traffic conditions of the dashed road are not part of the above traffic counts.

Pedestrians and bicycles influence the outcome of the simulation. Nevertheless, they are not observed for the calculation of the situation description for learning the classifier. The main function of these groups of road users is to increase the diversity of results in the training data and to model the most important types of road users occurring in reality.

Two classes are used to train a classifier.

$$class = \begin{cases} \text{congested} & \text{if } \bar{v} < 3.6 \text{ m} \cdot \text{s}^{-1} \\ \text{free} & \text{else} \end{cases}$$

We performed 20,000 simulation runs to train a classifier and 5,000 to measure its performance on unseen data. To define the settings of each simulation run, let $\mathcal{N}_a^b(\mu, \sigma) = \min(\max(\mathcal{N}(\mu, \sigma), a), b)$ be Gaussian distributed random number with μ and σ bounded to the interval $[a \cdots b]$. Table 1 shows the simulation parameters, calculated prior to each simulation run.

Figure 4 shows the distribution of training data. The data is separable and thus a classifier should be able to determine the classes.

The case scenario is a supervised learning scenario. The free machine learning software *WEKA*,[9] is used to analyze the training data and to build classifiers.

---

[9]http://www.cs.waikato.ac.nz/ml/weka/, see also Bouckaert et al. (2010).

Table 1 Different probabilities for each simulation run resulting in different volumes of traffic

$p_{create} = \mathcal{N}_{0.3}^{0.7}(0.5, 0.1)$	Probability to create a road user for each simulation iteration
$p_{car} = \mathcal{N}_{0.5}^{0.8}(0.75, 0.1)$	Probability that a created road user is a car
$p_{bicycle} = \mathcal{N}_{0.5}^{0.9}(0.75, 0.1)$	Probability, that it is a bicycle, when it is no car. Otherwise, it will be a pedestrian
$p_{A66} = \mathcal{N}_{0.67}^{0.9}(0.75, 0.1)$	Probability, that car will be put on the motorway A66
$p_{fixDest} = \mathcal{N}_{0.67}^{0.88}(0.75, 0.1)$	Probability to have one of the predefined destinations, otherwise the destination is a random point in the city
$p_{destInnerCity} = \mathcal{N}_{0.7}^{0.9}(0.8, 0.05)$	When having one of the predefined destinations: Probability to have the destination in the inner city of Frankfurt am Main, leading to use the dashed road. Otherwise the destination will be the P+R parking garage

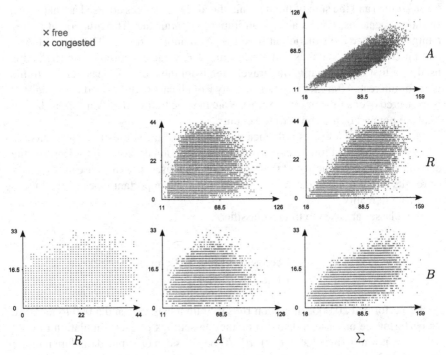

Fig. 4 Distribution of training data

For learning, we use WEKA's J4.8 implementation of the C4.5 algorithm (Quinlan 1993) and WEKA's JRip implementation of RIPPER (Cohen 1995), as they produce comprehensible classifiers.

The trained classifiers are shown in Figs. 5 and 6. RIPPER and C4.5 present the results in different ways. RIPPER gives one logical formula, presented in a summarized version and C4.5 generates a decision tree.

$$congested \; = \; (R \geq 22)$$
$$\vee \; (R \geq 19) \wedge (B \geq 10)$$
$$\vee \; (R \geq 17) \wedge (B \geq 15)$$
$$\vee \; (R \geq 21) \wedge (B \geq 6) \wedge (43 \leq A \leq 51)$$
$$\vee \; (R \geq 20) \wedge (B \geq 7) \wedge (58 \leq A \leq 64) \wedge (\Sigma \geq 72)$$
$$\vee \; (R \geq 16) \wedge (B \geq 8) \wedge (42 \leq A \leq 50) \wedge (S \leq 0) \wedge (72 \leq \Sigma \leq 75)$$
$$\vee \; (R \geq 14) \wedge (B \geq 12) \wedge (A \leq 62) \wedge (S \geq 1) \wedge (\Sigma \geq 91)$$
$$\vee \; (14 \leq R \leq 16) \wedge (11 \leq B \leq 17) \wedge (A \leq 50) \wedge (S \leq 1) \wedge (\Sigma \geq 78)$$

with $R = \#Riederwald$

$B = \# \, Bergen{-}Enkheim$

$A = \#A66$

$S = \# \, Seckbach$

$\Sigma = B + R + S + A$

Fig. 5 Learned decision rule using RIPPER

GIS-based Traffic Simulation using OSM

```
Riederwald <= 16: free
Riederwald > 16
|  Riederwald <= 22
|  |  Bergen-Enkheim <= 14
|  |  |  Riederwald <= 19
|  |  |  |  Bergen-Enkheim <= 9: free
|  |  |  |  Bergen-Enkheim > 9
|  |  |  |  |  Seckbach <= 1
|  |  |  |  |  |  Riederwald <= 18: free
|  |  |  |  |  |  Riederwald > 18
|  |  |  |  |  |  |  Bergen-Enkheim <= 13: free
|  |  |  |  |  |  |  Bergen-Enkheim > 13: congested
|  |  |  |  |  |  Seckbach > 1
|  |  |  |  |  |  |  Riederwald <= 18
|  |  |  |  |  |  |  |  A66 <= 50
|  |  |  |  |  |  |  |  |  sum <= 74
. . .
```

Fig. 6 Excerpt of the learned tree using C4.5

Both classifiers perform well on unseen data. Table 2 displays the confusion matrices of the learned classifier on the test set of 5,000 further simulation runs. The results are on par with each other: RIPPER achieves a predictive accuracy of 83.66% and C4.5 of 83.48%, respectively. The classifiers perform better than the simple solution to always choose "free" as a result, because it is the class with the highest frequency of occurrence, leading to a performance of 58.38%.

The small example shows the feasibility to use data mining techniques for traffic scenarios. In a next step, the cars coming from the motorway A66 and from Bergen-Enkheim could be advised to change destination to the P+R parking garage, when it is likely that a traffic jam will occur.

Table 2 Confusion matrices of the trained classifiers gained on the test set

RIPPER			C4.5		
Classified as →			Classified as →		
Correct result ↓	Congested	Free	Correct result ↓	Congested	Free
Congested	1,768	313	Congested	1,731	350
Free	504	2,415	Free	476	2,443
Correctly classified: 4,183 (83.66%)			Correctly classified: 4,174 (83.48%)		

7 Summary and Perspectives

An introduction on how to build up an executable traffic simulation from OSM cartographical material has been presented. The use of GIS technologies offers the opportunity to enhance the simulation model with data from various sources. For example, a digital terrain model could influence the routing of bicycles or a layer giving information about rain on the map could decrease the amount of bicycles and pedestrians in the simulation and make the cars drive more slowly.

The requirements for simulation of urban traffic have been discussed and three basic types of road users – car, bicycle and pedestrian – have been identified. Microscopic simulation models taking account for the interactions between these groups have been presented.

A case scenario has shown a method how to predict traffic jams in urban scenarios on the example of Frankfurt am Main. Machine learning techniques have been applied to train a classifier on the results of simulation runs. The classifier performs well on unseen data (approximately 83.5% correctly classified samples).

As one next step, a model for gas consumption and CO_2 emission can be applied to the car model as proposed in Dallmeyer et al. (2012b). Then, the influence of bicycles and pedestrians on this dimension could be analyzed. The influences of actions like advising to park the car in a P+R parking garage on traffic jams and on gas consumption then could be investigated.

Different types of classification algorithms could be compared in future work. It might also be advantageous to use different intervals than $\Delta = 300$ s.

Acknowledgements This work was made possible by the *MainCampus* scholarship of the *Stiftung Polytechnische Gesellschaft Frankfurt am Main*.

References

Agrawal R, Srikant R (1994) Fast algorithms for mining association rules. In: Proceedings of the 20th international conference on very large data bases, VLDB, Santiago de Chile, Sept 1994, pp 487–499

Bazzan ALC (2009) Opportunities for multiagent systems and multiagent reinforcement learning in traffic control. Auton Agents Multi-Agent Syst 18:342–375

Bennett J (2010) OpenStreetMap. Packt Publishing, Olton Birmingham, GBR. ISBN:978-1-84719-750-4

Blue VJ, Adler JL (2001) Cellular automata microsimulation for modeling bi-directional pedestrian walkways. Transp Res Part B Methodol 35(3):293–312

Bouckaert, RR Frank E, Hall MA, Holmes G, Pfahringer B, Reutemann P, Witten IH (2010) WEKA – experiences with a Java open-source project. J Mach Learn Res 11:2533–2541

Cohen WW (1995) Fast effective rule induction. In: Proceedings of the 12th international conference on machine learning, Lake Taho

Cormen TH, Stein C, Rivest RL, Leiserson CE (2001) Introduction to algorithms, 2nd edn. McGraw-Hill, New York

Dallmeyer J, Lattner AD, Timm IJ (2011) From GIS to mixed traffic simulation in Urban scenarios. In: Liu J, Quaglia F, Eidenbenz S, Gilmore S (eds) 4th international ICST conference on simulation tools and techniques, SIMUTools'11, Barcelona, 22–24 Mar 2011. ICST, Brüssel, pp 134–143. ISBN:978-1-936968-00-8

Dallmeyer J, Lattner AD, Timm IJ (2012a) Pedestrian simulation for urban traffic scenarios. In: Bruzzone AG (ed) Proceedings of the summer computer simulation conference 2012. 44rd summer simulation multi-conference, Genoa, 8–11 July 2012, S. 414–421. Curran Associates Inc

Dallmeyer J, Taubert C, Lattner AD, Timm IJ (2012b) Fuel consumption and emission modeling for urban scenarios. In: Troitzsch KG, Möhring M, Lotzmann U (eds) Proceedings of the 26th European conference on modelling and simulation (ECMS 2012), Koblenz, pp 574–580

Fellendorf M (1994) Vissim: a microscopic simulation tool to evaluate actuated signal control including bus priority. In: 64th institute of transportation engineers annual meeting, Dallas

Hafstein SF, Pottmeier A, Wahle J, Schreckenberg M (2003) Cellular automaton modeling of the autobahn traffic in north rhine-westphalia. In: Proceedings of the 4th MATHMOD, Vienna, pp 1322–1331

Haklay M (2010) How good is volunteered geographical information? A comparative study of OpenStreetMap and ordnance survey datasets. Environ Plan B Plan Des 37(4):682–703

Helbing D (1997) Empirical traffic data and their implications for traffic modeling. Phys Rev E 55(1):R25–R28

Helbing D (2001) Traffic and related self-driven many-particle systems. Rev Mod Phys 73(4):1067–1141

Ishaque MM, Noland RB (2008) Behavioural issues in pedestrian speed choice and street crossing behaviour: a review. Transp Rev Transnatl Transdiscipl J 28(1):61–85

Jain AK, Murty MN, Flynn PJ (1999) Data clustering: a review. ACM Comput Surv 31(3):264–323

Johnson M, Newstead S, Charlton J, Oxley J (2011) Riding through red lights: the rate, characteristics and risk factors of non-compliant urban commuter cyclists. Accid Anal Prev 43(1):323–328

Knospe W, Santen L, Schadschneider A, Schreckenberg M (2002) A realistic two-lane traffic model for highway traffic. J Phys A Math Gen 35(15):3369–3388

Krauss S, Wagner P, Gawron C (1996) Continuous limit of the nagel-schreckenberg model. Phys Rev E 54(4):3707–3712

Larsen J, El-Geneidy A (2011) A travel behavior analysis of urban cycling facilities in Montréal, Canada. Transp Res Part D Transp Environ 16(2):172–177

Lattner AD, Dallmeyer J, Timm IJ (2011) Learning dynamic adaptation strategies in agent-based traffic simulation experiments. In: Klügl F, Ossowski S (eds) Ninth German conference on multi-agent system technologies (MATES 2011). LNCS 6973. Springer, Berlin, pp 77–88

Nagel K, Schreckenberg M (1992) A cellular automaton model for freeway traffic. J Phys I 2(12):2221–2229

Nagel K, Beckman RJ, Barrett CL (1999) Transims for urban planning. Los Alamos unclassified report LA-UR 98-4389

Ottomanelli M, Caggiani L, Giuseppe I, Sassanelli D (2009) An adaptive neuro-fuzzy inference system for simulation of pedestrians behaviour at unsignalized roadway crossings. In: 14th online world conference on soft computing in industrial application, 14. http://link.springer.com/content/pdf/10.1007%2F978-3-642-11282-9_27.pdf

Pei J, Han J, Mortazavi-Asl B, Pinto H, Chen Q, Dayal U, Hsu M-C (2001) PrefixSpan: mining sequential patterns efficiently by prefix-projected pattern growth. In: Proceedings of the 12th IEEE international conference on data engineering, Heidelberg, pp 215–224

Quinlan JR (1993) C4.5 – programs for machine learning. Morgan Kaufmann, San Mateo

Rieser M (2010) Adding transit to an agent-based transportation simulation concepts and implementation. Dissertation, Technische Universität Berlin

Russell S, Norvig P (2003) Artificial intelligence: a modern approach, 2nd edn. Prentice Hall/Pearson Education, Upper Saddle River

Zielstra D, Zipf A (2010) A comparative study of proprietary geodata and volunteered geographic information for Germany. In: 13th AGILE international conference on geographic information science 2010, Guimarães

Zilske M, Neumann A, Nagel K (2011) OpenStreetMap for traffic simulation. In: State of the map Europe (SOTM-EU), Vienna

Evaluation of Real-Time Traffic Applications Based on Data Stream Mining

Sandra Geisler and Christoph Quix

Abstract Traffic management today requires the analysis of a huge amount of data in real-time in order to provide current information about the traffic state or hazards to road users and traffic control authorities. Modern cars are equipped with several sensors which can produce useful data for the analysis of traffic situations. Using mobile communication technologies, such data can be integrated and aggregated from several cars which enables intelligent transportation systems (ITS) to monitor the traffic state in a large area at relatively low costs. However, processing and analyzing data poses numerous challenges for data management solutions in such systems. Real-time analysis with high accuracy and confidence is one important requirement in this context. We present a summary of our work on a comprehensive evaluation framework for data stream-based ITS. The goal of the framework is to identify appropriate configurations for ITS and to evaluate different mining methods for data analysis. The framework consists of a traffic simulation software, a data stream management system, utilizes data stream mining algorithms, and provides a flexible ontology-based component for data quality monitoring during data stream processing. The work has been done in the context of a project on Car-To-X communication using mobile communication networks. The results give some interesting insights for the setup and configuration of traffic information systems that use Car-To-X messages as primary source for deriving traffic information and also point out challenges for data stream management and data stream mining.

Keywords Data streams • Data stream mining • Data quality • Traffic management • Intelligent transportation systems

S. Geisler (✉) • C. Quix
Information Systems, RWTH Aachen University, Ahornstr. 55, 52056 Aachen, Germany
e-mail: geisler@dbis.rwth-aachen.de; quix@dbis.rwth-aachen.de

G. Cervone et al. (eds.), *Data Mining for Geoinformatics: Methods and Applications*, 83
DOI 10.1007/978-1-4614-7669-6_5, © Springer Science+Business Media New York 2014

1 Introduction

When you buy a new car today, it is usually equipped with a multitude of sensors. The sensors have mainly two purposes: increasing safety and increasing comfort. Driver assistance systems, such as brake, lane, or parking assistants, are realized with these sensors. The substantial progress in making road transport safer resulted in a continuously decrease of accidents involving personal injuries and road fatalities in Germany – the number of road fatalities being lower than ever since 1950 (Statistisches Bundesamt 2011). A main goal to improve safety further, should be to not only lower the severity of accidents, but to prevent them to the greatest possible extent. This can be done by prevention mechanisms which warn road users before they approach a critical situation. To this end, communication between vehicles and their surroundings, e.g., other vehicles or traffic infrastructure, termed Car-to-X Communication (C2X), can help to exchange important safety information "sensed" by vehicles. In the Cooperative Cars (CoCar) project and its successor Cooperative Cars eXtended (CoCarX),[1] which constitutes the context of our work, a heterogeneous approach has been followed. Depending on the use case, 802.11p-based technology (pWLAN) or UMTS and its successor LTE are used for communication (cf. Fig. 1). In the CoCar system, hazard warnings are sent by vehicles to other road users. For example, when a vehicle brakes very hard, a warning message is sent to vehicles in vicinity (also called *Geocasting* or *GeoMessaging* (Fiege et al. 2011)).

Our goal in the CoCar project is to utilize the sensor information included in the messages of vehicles for deriving higher value information, such as the position of a queue-end or the current traffic state. However, there are several challenges in doing so. First, the amount of messages per time sent by vehicles from a specific area can be very high. Second, many traffic applications, especially safety applications, and their users require real-time processing and response. Data Stream Management

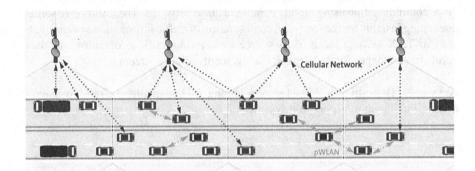

Fig. 1 The CoCarX architecture (Fiege et al. 2011)

---

[1]http://www.aktiv-online.org/english/aktiv-cocar.html

Systems (DSMS) and data stream mining algorithms proved to be suitable for several application fields with real-time properties. Also in traffic applications, data stream processing and mining has been applied successfully (Kargupta et al. 2010; Biem et al. 2010; Ali et al. 2010; Arasu et al. 2004), but a systematic approach for evaluating the parameters and results of such applications is still missing. Especially, the comparison of multiple data stream mining algorithms according to their suitability for various ITS applications is an interesting parameter to be analyzed systematically.

Furthermore, the quality of the processed data and the produced information is crucial. Users want to rate the reliability of information, and low-quality information should not be distributed to users to avoid desensitization and frustration. This applies not only to traffic information systems, but also to other geographic applications which rely on sensor data and apply data mining techniques to derive new information (e.g., tsunami or earthquake warning systems). The data in such applications is inherently uncertain as sensor data might be inaccurate. In addition, the methods for data analysis usually cannot achieve a 100% accuracy. Hence, the continuous quality monitoring of the processed data should be an inherent property of a data management solution for such applications.

In this paper we will present a summary of our work on a data stream-based evaluation framework for traffic applications. We first introduce the overall architecture of the evaluation framework in Sect. 2, followed by the description of the data processing and the corresponding data quality processing in Sect. 3. Subsequently, two case studies in queue-end detection and traffic state estimation are detailed in Sect. 4. The section contains also the evaluation results for various configurations of the ITS and multiple data stream mining algorithms. In Sect. 5, we will compare our approach with other works from related research areas. Finally, the lessons we learned throughout the project and some ideas for future work are discussed in Sect. 6.

2 Architecture

The evaluation framework is based on a data management and fusion architecture which was designed and implemented in the first phase of the CoCar project (Geisler et al. 2009). The main constituents of the architecture are mobile data sources which deliver the raw sensor data. Second, the data is received by a Data Stream Management System (DSMS) which processes and integrates the data. Third, to analyze the data and derive new traffic information, we integrated a data stream mining framework into the DSMS. In addition, we use a spatial database to speed up working with geospatial data. We detail these components in the following. Finally, we added a data quality monitoring component to the architecture. An overview of the architecture with its main constituents is depicted in Fig. 2.

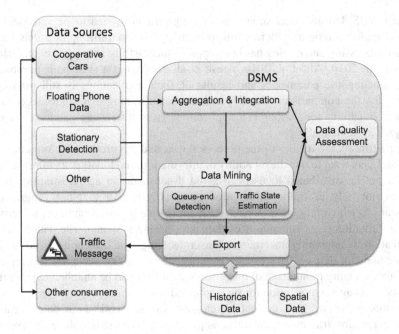

Fig. 2 The overall architecture

2.1 Mobile Data Sources

For reasons of reproducibility, controllability, and creation of a sufficient amount of data, we use the traffic simulation software VISSIM by PTV AG[2] to simulate the creation of data from mobile and stationary sensors. In the CoCarX project, one important data source is, of course, the CoCars sending event-based messages in case of hazards to warn other vehicles. We create CoCar messages by using the VISSIM simulation for the following events: a vehicle braking very hard and a vehicle turning on its warning flashers. For traffic state estimation we introduced another type of message, which is sent periodically by the vehicles and only contains basic information, such as position and speed. The second important data source is Floating Phone Data (FPD). FPD are anonymously collected positions of mobile phones, but the location data might be very inaccurate. Other traffic data, such as the vehicle speed, can be derived from FPD (Geisler et al. 2010). In our case studies, we focus on the CoCar messages only as we want to analyze the usefulness of CoCar messages and the influence of parameters, such as required equipment rates, for certain traffic applications.

---

[2]http://www.ptv.de

2.2 Data Stream Management System

We use the Global Sensor Network (GSN) system (Aberer et al. 2006) for data stream management. GSN provides a flexible, adaptable, distributable, and easy to use infrastructure. It wraps the functionality required for data stream processing and querying around existing relational database management systems, such as MySQL. The main concepts in GSN are wrappers and virtual sensors. In GSN, we provide a wrapper for each data source to receive the data. Each further processing step is encapsulated into a virtual sensor which creates the output data by a query over the input data. For data stream mining, we integrated the data stream mining framework Massive Online Analysis (MOA) (Bifet et al. 2010) as a virtual sensor into the DSMS. The DSMS distributes the derived information to consumers, e.g., as queue-end warning messages to CoCar vehicles.

2.3 Data Stream Mining

As mentioned in the previous paragraph, a virtual sensor has been built which classifies the aggregated data, i.e., each data stream element. We started off with integrating the MOA framework and all of the results presented in Sect. 4 have been made using this framework. We also made this virtual sensor more flexible, such that different kinds of data (stream) mining algorithms can be used in a convenient way. The MOA framework offers several algorithms, mainly based on Very Fast Decision Trees, a variant of the Hoeffding Tree, first proposed by Domingos and Hulten (2000). Hoeffding trees use a statistical measure called Hoeffding bound. The bound determines, how many examples are required at each node to choose a split attribute. It guarantees, that with a given probability the same split attribute is selected as the attribute which would have been chosen if all training elements were known already. The algorithm first determines the two best attributes by using a relevance measure such as information gain or the Gini index. Afterwards, if the difference between the two attributes' relevance values excels the calculated Hoeffding bound for the examples seen so far, the best attribute is chosen for the split. For classification, the MOA framework provides single learners with various voting strategies as well as ensemble algorithms and concept-adapting algorithms. For detailed descriptions of the algorithms please see Bifet and Kirkby (2009). In the mining virtual sensor, the classified stream elements are extended with an attribute for the classification result, e.g., the estimated traffic state.

2.4 Spatial Database System

Spatial database systems have been introduced to ease and speed up the work with spatial data using special data types and functions. In our architecture, we store and query the road network by using the spatial functionality of Microsoft SQL Server

2008.[3] We export the roads (or links) from the traffic simulation and store them as curve objects in the database. In the database, we divide the links into sections of equal length, e.g., sections of 100 m length each. Sections are the units of interest for our evaluations, e.g., the traffic state will be determined for each section. Each time we have to locate a vehicle on the network, the spatial database is queried to find the section the vehicle is driving on.

3 Data Stream Processing

To show the effectiveness of our framework for the purpose of a systematic evaluation of ITS applications, we set up two case studies, namely queue-end detection and traffic state estimation, detailed in Sect. 4. We will first explain the processing of the data streams throughout the DSMS as it is similar for both case studies. Subsequently, the data quality framework we implemented to measure and rate the effectiveness and efficiency of the traffic applications is elucidated.

3.1 Processing Flow

GSN provides two structures to work with data streams: wrappers and virtual sensors. The wrappers manage the connections to the data sources, receive the data and convert it to relational stream elements. In our case studies, the CoCar messages and ground truth messages (stating the real queue-end and the real traffic state) are produced by the traffic simulation and sent via TCP to the DSMS. Virtual sensors filter and process the data, e.g., aggregate and integrate data. One common issue in mobile data detection is that the measured positions contain some error whose amount depends on the used positioning technique. This error influences the accuracy of traffic information (Geisler et al. 2010). In case of the CoCar messages, we can assume GPS positioning accuracy. To approximate reality as close as possible, the exact positions in the messages created by VISSIM have to be degraded. To this end, the stream elements created from the CoCar messages are forwarded to a degradation virtual sensor. The next virtual sensor matches the positions of the CoCar messages to the road network, a process also termed Map Matching. It tries to find the section where the vehicle actually is located.

The next step is the preparation of the data for data stream mining. In our approach, we aggregate data, such as speed or acceleration, of CoCar messages coming from the same section for a certain time window. We utilize classification algorithms for data stream mining in our framework, i.e., each incoming tuple is

---

[3]Other products can also be used in the architecture. We implemented the same functionality also for PostgreSQL and PostGIS.

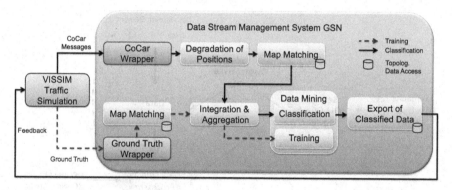

Fig. 3 Data stream processing flow

assigned to a specific class. For example, in queue-end detection either the class "Queue End" or the class "No Queue End" will be assigned to each stream element. For each run a classifier was learned from scratch. Afterwards, the classified element is distributed to consumers, e.g., sent to the traffic simulation, where it can be used to visualize the corresponding event. The message contains a confidence value, which indicates the reliability of the message based on data quality parameters calculated throughout the data processing in the DSMS. The complete data stream processing flow is depicted in Fig. 3.

3.2 Data Quality Processing

Data quality (DQ) plays an important role in DSMS as there is usually a trade-off between accuracy and consistency on the one hand, and timeliness and completeness on the other hand. We follow a holistic approach for DQ management in data streams which is based on a comprehensive ontology-based data quality framework. We will briefly explain the framework here, as we use it to measure the effectiveness and efficiency of the traffic applications in our experiments with it. The framework is presented in detail in Geisler et al. (2011).

Our aims in designing a DQ framework for DSMS were flexibility and adaptivity as the framework should not be restricted to traffic management applications. Even within the traffic domain, there are different ways to measure DQ depending on the data sources which are available, and the data processing steps which are applied in the DSMS. For example, to measure the correctness of a CoCar message which signals a specific event (e.g., an icy road has been detected by the car sensors), we may check whether this information is confirmed also by other cars in the same area. This can be done by computing a correctness measure while several CoCar messages are aggregated in a query in the DSMS. If an external source with weather information is available, we may also want to apply a rule to check the consistency of the information (e.g., if the temperature is above 10° C, the road cannot be icy).

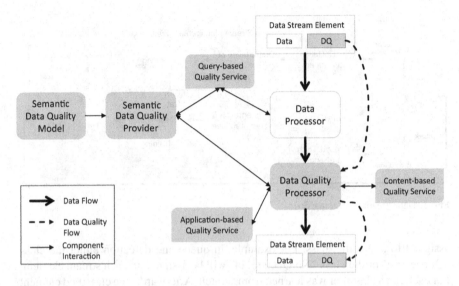

Fig. 4 Architecture of the ontology-based DQ component

Therefore, it is not reasonable to just select a set of DQ dimensions and DQ metrics, and hardcode them in the DSMS. A flexible and extensible DQ framework is required which allows also the definition of application-specific DQ metrics. In our approach, we distinguish three types of DQ metrics: (i) *content-based* metrics, (ii) *query-based* metrics, and (iii) *application-based* metrics. Content-based metrics use semantic rules and functions to measure data quality, e.g., as stated in the consistency check mentioned above. Query-based metrics are methods to compute the DQ for certain query operators, e.g., the accuracy of an aggregation operator is the average accuracy of the input tuples. We use a query rewriting approach to offer more flexibility than in other approaches (e.g., Abadi et al. 2005; Klein and Lehner 2009). Finally, application-based metrics measure DQ by any kind of application-specific functions, e.g., confidence values of a data mining classifier are provided by the data mining system.

Architecture of the DQ Framework

The data quality framework consists of several components, which are depicted in Fig. 4. To enable a flexible approach, all data quality related metadata is organized in a semantic data quality model, namely a data quality ontology. An ontology is a well suited tool to formalize the DQ model as it can be easily extended and adapted by users due to its human-readability and availability. The ontology includes a data stream part, which describes the concepts of data streaming, such as streams, windows, tuples, or attributes. Furthermore, concepts describing data quality, such as data quality metrics or dimensions, are incorporated in a data quality part. The data quality factor concept establishes a relationship between these parts defining which quality dimensions are measured with which metric and for which stream element in the data stream management system.

When incorporating DQ into the DSMS, the set of data stream attributes is extended by a set of DQ attributes, where each attribute represents a DQ dimension. We define these streams as *quality-affine* data streams. Syntactically, quality attributes are not distinguishable from data attributes, but semantically, they are handled differently and must be recognizable during data processing.

The three different types of DQ metrics are supported in the framework by corresponding *quality services*. The *Semantic Data Quality Model* is the DQ ontology presented before. The *Semantic Quality Provider* acts as an interface to this ontology. The *DQ Processor* invokes the *Content-based* and *Application-based Quality Services* as required by the definitions in the ontology. The *Query-Based Quality Service* interacts directly with the *Semantic Data Quality Provider* and the data processor, as queries have to be modified in the data processing steps. In Fig. 4 all grey boxes are DQ components and are added as additional components to the DSMS.

To take into account DQ processing, the semantic descriptions for the DQ assessment are loaded from the ontology at system startup. Based on this information, the virtual sensor configurations are rewritten recursively, including the queries and the output structure of each virtual sensor in the dependency graph.

We have used the DQ component in our case studies to measure important means to rate the effectiveness and efficiency of a traffic application. For example, we measured the data mining accuracy for varying parameters and the processing time of the stream elements in the system. For the case studies described in Sect. 4, additional quality factors have been defined which are combined into a final DQ value that represents the confidence in the derived information (how reliable is the information, that a queue end has been detected on a specific section?). We showed in Geisler et al. (2011), that the quality values and confidence values produced by the data quality component indeed indicate a drop or increase in data quality (e.g., the accuracy of the Map Matching using mobile phone positions is lower than using GPS positions). We could also prove that the data quality component produces only a slight overhead in performance, memory, and CPU consumption (Geisler et al. 2011).

4 Case Studies

To show the effectiveness of our framework for the purpose of systematic evaluation of ITS applications, we set up two case studies. In this section, we summarize the experiments and results of the queue-end detection and the traffic state estimation use cases. Further details about the two case studies and results can be found in Geisler et al. (2012).

Fig. 5 Real and estimated queue-ends in the traffic simulation. The *dark gray vehicles* send a CoCar message in this time step

4.1 Queue-End Detection

The queue-end detection scenario aims at the detection of hazards caused by traffic congestion. The end of a queue constitutes a threat, as drivers may not be prepared to suddenly approach a static obstacle. The aim of our experiments was to investigate the influence of multiple parameters on the detection accuracy. For each section of the roads in the network it is determined, if it contains a queue-end or not. In this case study, traffic has been simulated on a simple road network consisting of a highway of 5–10 km length, i.e., a road with two directions with two lanes each.

One direction contains a hazard (a construction site with an excavator) narrowing the street to one lane. Depending on the traffic volume, the hazard causes a congestion. The traffic simulation sends the CoCar messages produced during simulation runs and periodic information about the ground truth (i.e., positions of queue-ends) to the DSMS. The DSMS processes this information as described in the previous section. The mining algorithm first decides, based on the received information for one section, if it contains a queue-end (classification) and learns, in which cases a queue-end is present in a section (training). If the classification identified a queue-end, a hazard warning is sent to the traffic simulation and is visualized by a red (estimated position) traffic sign as depicted in Fig. 5. The blue traffic sign denotes the queue-end as estimated by the traffic simulation (based on the maximum queue length in the last 10 s). In our simulation the queue-end warnings were located in the middle of the corresponding section. In a real-world scenario this should be set to the end of the section for safety reasons.

We tested the influence of several parameters on the output of the classification. These parameters can be divided into two categories: parameters inherent to the traffic situation and parameters inherent to data processing. Traffic situation parameters include the percentage of Cooperative Cars in the overall vehicles (penetration rate, default value: 5%), the traffic volume (vehicles per hour, default value: 3,000 veh/h), and the section size (default value: 100 m). In each simulation run (excluding only the runs with the variation of traffic volume), exactly the same traffic situation is reproduced; thus, the results of different runs are comparable. Data processing parameters comprise the size of the sliding time window for data aggregation (default value: 120 s), the mining algorithm (default algorithm: Hoeffding Tree), and the ratio of positive (section contains queue-end) and negative (section contains no queue-end) training examples presented to the mining algorithm (default option: as ratio appears, no balancing of examples). When investigating one of the before mentioned six parameters, only this parameter has been varied, all other parameters kept their default value.

For each experiment consisting of one simulation run, the following measures have been recorded: the overall accuracy (ratio of correctly classified elements to all classified elements), the sensitivity (or recall, ratio of correctly determined positive examples to all positive examples), precision (ratio of all correctly determined positive examples to the sum of all correct positive and all false positive examples), and the specificity (ratio of correctly determined negative examples to all negative examples). We recorded and plotted these measures (y-axis, each in percentage) over simulation time (x-axis, in seconds) to rate the learning progress of the algorithms during the simulation runs. If not otherwise stated, the algorithms build models from scratch in each run.

For this case study, each stream element produced by the aggregation sensor and processed for training by a mining algorithm, has the following structure:

$$MiningElement(Timed, AvgSpeed, AvgAccel, HasQueueEnd,$$

$$WLANo, EBLNo, LinkID, SectionID)$$

where `Timed` is the timestamp assigned by the aggregation sensor as creation time of the stream element, `AvgSpeed` is the average of speeds from the CoCar messages of the last time window (e.g., the last 120 s) for the section and link at hand, `AvgAccel` is the average acceleration from the CoCar messages for the same window, section and link, `HasQueueEnd` is the ground truth for this section (does it contain a queue-end or not in the traffic simulation?), `WLANo` is the number of messages of the type warning light announcement, `EBLNo` the number of emergency braking light messages for this section and time window, `LinkID` is the id of the link, and finally `SectionID` is the id of the section which the data was aggregated for. An example stream element would be:

$$(1338981779163, 19.5, -6.75, 1, 0, 3, 21, 2)$$

The output stream elements of the mining sensor have almost the same structure, containing additional fields for data quality, such as the mining accuracy or the timeliness, and of course the class estimated by the algorithm:

$$MiningElement(Timed, AvgSpeed, AvgAccel, HasQueueEnd,$$

$$WLANo, EBLNo, LinkID, SectionID,$$

$$ClassQueueEnd, DQ\_Timeliness, DQ\_MiningAccuracy)$$

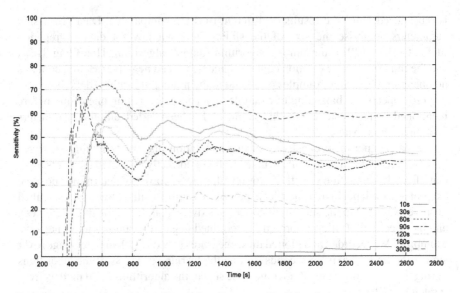

Fig. 6 Sensitivity results for varying window sizes over simulation time

We identified a set of default values for all parameters, and while one parameter was varied in experiments, all other parameters kept their default value. In most experiments, the mining accuracy was dominated by the specificity, as we had far more negative than positive examples.

In the experiments with varying window sizes between 10 and 300 s, we identified a default window size of 120 s as a good compromise for getting acceptable results for sensitivity (about 51%) as well as for specificity (about 92%), and precision (about 61%). The results for sensitivity are depicted in Fig. 6. Simulation runs with varying traffic volumes ranging from 2,000 to 4,000 showed, that, though the number of messages increases and the sensitivity is better for higher traffic volumes, it always deteriorates in the end. This is due to the fact that the ratio of positives to negatives gets highly imbalanced (ranging from 14% in 2,000 veh/h run to 9% in 4,000 veh/h run), because there are many sections that do not include a queue-end but deliver many messages. Precision on the other hand, seems to be best for 3,000 veh/h, but it does not show a clear trend for increasing or decreasing traffic volumes.

For the test with multiple penetration rates (from 1 to 10%, cf. Fig. 7) we found out, that the accuracy is not substantially influenced by the penetration rate, although higher penetration rates (from 5% on) deliver a slightly better sensitivity than the lower values. In contrast, precision increases with increasing penetration rates. For 10% penetration rate, precision gets up to 88% precision as shown in Fig. 8.

The results for simulation runs with section lengths between 30 and 300 m showed that the sensitivity increases with increasing section length. However, more tests have to be done to see if it reaches a maximum, where it decreases again. For Precision on the other hand, the best values are reached for 150 m with no clear trend with increasing or decreasing section lengths (30 m being the worst).

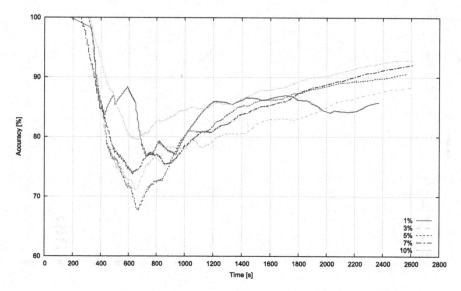

Fig. 7 Accuracy results for varying penetration rates over simulation time

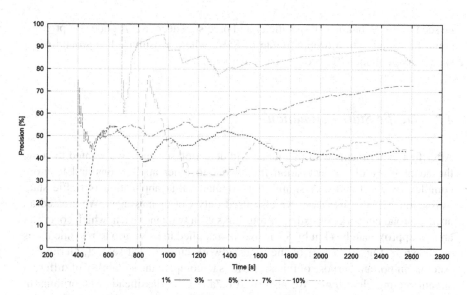

Fig. 8 Precision for varying penetration rates over simulation time

To tackle the problem of the unbalanced ratio between positive and negative examples, we also made tests using an undersampling technique in the data mining process. It drops negative examples randomly; we experimented with percentages between 0 and 90%. The sensitivity increases with increasing dropping rate as can

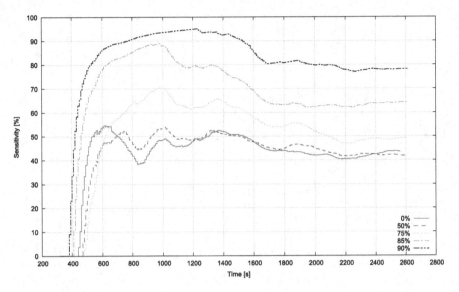

Fig. 9 Sensitivity for multiple undersampling rates over simulation time

be seen in Fig. 9, but in turn specificity and precision decrease. A good compromise could be found with a dropping rate of 85%, where sensitivity lies around 65%, specificity at about 82%, and precision at about 38%.

4.2   Traffic State Estimation

After first experiments with the queue-end detection scenario, we wanted to show the adaptability of the evaluation framework to other applications and to more complex road networks. A simple but popular traffic application is traffic state estimation. For this use case, we also used only CoCar messages. We created an artificial road network consisting of four links with two lanes each, which covers an area of approximately 4 km by 8 km. We also applied the scenario to realistic maps as shown in Fig. 10 (a part of the road network close to Düsseldorf, Germany). Another important purpose of this scenario is to compare the suitability of different stream mining algorithms. We also used data stream classification algorithms in the experiments: data stream elements are classified into one of four traffic states (free, dense, slow-moving, and congested as proposed in BASt (1999)). The ground truth for each section is calculated in the traffic simulation using the corresponding rules defined in BASt (1999). To evaluate the results, we compiled a confusion matrix which indicates the combinations of real classes and estimated classes (e.g., number of elements whose real class is dense, but have been classified as free). Based on the confusion matrix the mining accuracy has been calculated. To show the representativeness of our results, we made three runs varying only the seed

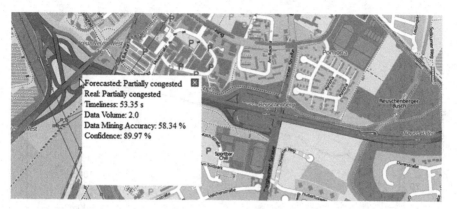

Forecasted: Partially congested
Real: Partially congested
Timeliness: 53.35 s
Data Volume: 2.0
Data Mining Accuracy: 58.34 %
Confidence: 89.97 %

Fig. 10 Visualization of the traffic state estimation scenario with OpenStreetMap

for random functions in the traffic simulation (e.g., used for vehicle speeds, entrance of vehicles in the simulation and so on). The comparison of these runs showed, that the overall accuracy of the mining algorithm only varied about 3–5% max.

The results are visualized on a map, where the estimated traffic state for each section is color-coded as shown in Fig. 10.

In the simulation runs we also used a set of default values for the variable parameters and a default classifier. We used 400 m for the section length, a window size of 60 s, and a 10% CoCar penetration rate as default values. The concept-adapting Hoeffding Option Tree with Naïve Bayes prediction strategy was used as the default classifier as it delivered the best results in comparison to other algorithms. We used no balancing algorithm in all the experiments. The input stream elements for the mining algorithms look similar to the queue-end detection scenario – only the ground truth attribute differed, containing the traffic states calculated in the traffic simulation as described. The output stream elements contain besides the additional estimated traffic state class, the timeliness, and the number of elements classified into each of the confusion matrix cells.

We first only used the emergency braking and warning light announcement messages for queue-end detection. However, we noticed that all traffic states except for congested were not detected well, since the event messages are only created in crowded traffic situations. Hence, we introduced periodically sent messages in the simulation which only contained basic non-event data (mainly position, speed, and acceleration). This improved the accuracy of the other classes, but still did not lead to good results, especially for the class dense. One problem seemed to be again the imbalanced number of examples for each class. Therefore, multiple traffic volumes on the different roads have been used in the traffic simulation to create training examples distributed over the classes (between 2,500 and 3,500 veh/h). Another improvement was the inclusion of the number of periodic messages as input parameter into the classified stream elements, which we assumed to be an indicator for the traffic density. This led to a substantial increase for the classes

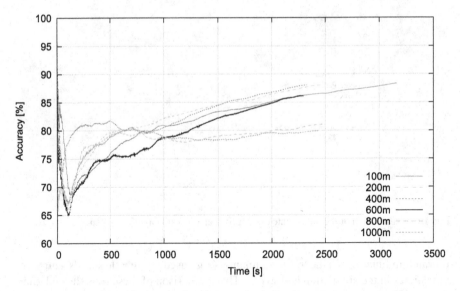

Fig. 11 Accuracy for varying section lengths over simulation time

free, dense, and slow-moving. Experiments with varying section lengths revealed, that sections with sizes of 100 m and over 600 m produce less accurate results than sizes of 200 and 400 m (around 87% overall accuracy). Smaller sections may not contain enough messages to indicate the traffic state, while larger sections can include more than one traffic state as depicted in Fig. 11. Hence, we used 400 m as a default value for the section length in further tests.

Similar to the queue-end detection scenario tests with varying penetration rates showed no substantial influence on the overall accuracy.

The comparison of the data stream mining algorithms of the MOA framework (Bifet et al. 2010) was divided into several experiments. First, we were interested in the influence of different voting strategies with the same classifier (in this case the basic Hoeffing Tree). It turned out that on average, the Naïve Bayes and the Adaptive Hybrid strategy were very similar in their results, but both outperformed the Majority Voting strategy. In the next group of experiments, the ensemble mining algorithms of the MOA framework were compared using the Hoeffding Tree as a base learner. The boosting algorithm (Oza 2005) was slightly better than the other algorithms with accuracies of up to 92%. However, it delivers also only a slightly better accuracy than the best single classifier algorithm in average. The results are summarized in Fig. 12.

Finally, also algorithms which are able to adapt to concept drift have been compared to see, if they can compete with the other algorithms (though our scenario does not contain concept drifts). The experiments showed, that none of them outperforms the ensemble algorithms, in fact the online boosting algorithm (Oza 2005) has proven to be slightly better than the concept drift-adapting variants.

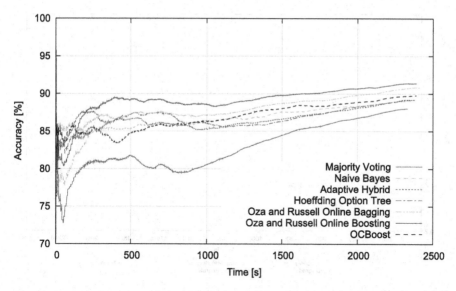

Fig. 12 Accuracy for various prediction strategies and ensemble algorithms over simulation time

Performance of data processing is a crucial aspect in data stream management systems as a huge amount of data has to be processed in a short time. Hence, we measured the processing time of the data stream elements for each sensor using the timeliness dimension. This means, that for each CoCar message the time between its creation in the traffic simulation and the processing time in the sensor is calculated. When data is aggregated over a window, the timeliness of the oldest stream element in the window is used as the timeliness value. Figure 13 shows the timeliness over simulation time for the Düsseldorf case study smoothed with a Bézier function. In this performance experiment in a 40 min simulation run, about 19,000 CoCar messages have been processed, aggregated to nearly 50,000 data stream elements which have been subsequently mined by a data stream mining algorithm. It can be seen from Fig. 13 that the sensors before aggregation only need a few milliseconds to process the stream elements, where the processing time is slowly increasing with simulation time. In the aggregation sensor, data of the last 120 s has been aggregated, hence the timeliness values are higher per se.

5 Related Work

Our work is related to a multitude of areas. DSMS have already been used to realize real-time traffic applications (Biem et al. 2010; Ali et al. 2010; Arasu et al. 2004; Kargupta et al. 2010), but none of them provided a thorough analysis of the applications, their efficiency, and the influencing factors to derive advices

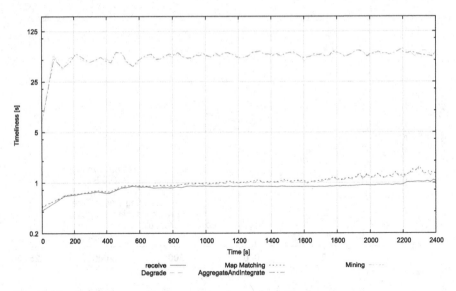

Fig. 13 Timeliness of stream elements measured over simulation time

for traffic information providers. An evaluation framework for traffic applications based on VISSIM has been proposed for example by Fontaine and Smith (2007). They investigated the influence of different parameters in the road network and mobile network on the accuracy of traffic information, such as link speed. However, they do not consider the characteristics of the processing information system and do not incorporate analysis techniques, such as data mining, to derive events or traffic information from the collected data. In our work, we investigate the influence of varying road and traffic parameters, and we also take into account the special features of the DSMS (e.g., window size, sliding step) and of the stream mining algorithms, such as concept drift. For queue-end detection, Khan presents a simulation study also based on VISSIM using stationary sensors as data source (Khan 2007). The data is analyzed using Artificial Neural Network models, predicting the current queue length based on accumulated numbers of cars and trucks at fixed locations. Although they postulate real-time processing in their information system, the used algorithm is not capable of online learning, i.e., it cannot adapt to concept changes, which can be achieved by using data stream mining algorithms as proposed in our approach. Neural networks have also been used for traffic state estimation and short-term prediction (Bogenberger et al. 2003; Chen and Grant-Muller 2001). As neural networks are easily applicable for data streams, we will also compare some of them with the performance of the decision tree algorithms included in MOA in future work.

Approaches for data quality extensions in DSMS mainly consider Quality of Service aspects, i.e., to monitor the system performance during query processing and adapt the system configuration if certain quality criteria are not met. Some of these approaches have a very coarse granularity and are only considering the quality of the outputs (e.g., Schmidt et al. 2004; Abadi et al. 2003), while others measure data quality at each operator and for each stream element or window (Abadi et al. 2005; Klein and Lehner 2009). Most approaches allow system-based quality measures, such as throughput or performance, while some also provide user-defined and content-related quality measurement beyond Quality-of-Service applications, e.g. in Klein and Lehner (2009). In contrast, we allow a very flexible and fine-granular definition of data quality and its measurement in our framework. The framework is based on a modularized ontology which allows for defining system-based as well as application-based data quality metrics using rules and numerical expressions (Geisler et al. 2011).

6 Conclusion and Outlook

For intelligent transportation systems, and especially safety applications, high effectiveness is a crucial success factor. Applications producing low quality results are quickly rejected by users, because they do not trust them or are desensitized by false alarms. The presented framework is a step towards evaluating traffic applications based on mobile and static data sources in a realistic setting. The framework enables ITS providers to test several system configurations, data sources, and algorithms and allows to derive advices for productive systems. For example, we showed, that data from C2X messages can be utilized to derive hazard warnings even if the penetration rate is low. Of course, simulation studies always have to be validated against reality and one future aim is to try out the optimal set of parameter values in a real traffic scenario. For this, a model has to be learned in several simulation runs with varying traffic situations. The model can then be used for classification of data produced by real traffic situations. Depending on the effectiveness, the model has to be adapted in simulation again. To approximate reality in the simulation as close as possible, the real traffic data (e.g., speed measures) can be used in the traffic simulation.

Our work also revealed, that ITS are an interesting application area for data stream concepts and management systems. As current and future traffic applications and their users require real-time processing and response, data stream management and mining concepts proved to be a perfect match in the case studies we analyzed and look promising for other applications. Additionally, the continuous quality monitoring in the framework helps to keep up user trust and gives feedback to ITS developers. So far, we utilized existing processing and analysis algorithms to realize the described use cases, but the experiments also showed that the potential for tweaking has not been fully exploited. For example, we will investigate more complex techniques better suited for data streams to balance training examples. Example scenarios for concept drift, e.g., to simulate a full day with bursty and low traffic periods, would be also a challenge.

Acknowledgements This work has been supported by the German Federal Ministry of Education and Research (BMBF) under the grant 01BU0915 (Project Cooperative Cars eXtended, http://www.aktiv-online.org/english/aktiv-cocar.html) and by the Research Cluster on Ultra High-Speed Mobile Information and Communication UMIC (http://www.umic.rwth-aachen.de). We thank the PTV AG for kindly providing us with a VISSIM license. We also thank the reviewers for their valuable comments.

References

Abadi DJ, Carney D, Çetintemel U, Cherniack M, Convey C, Lee S, Stonebraker M, Tatbul N, Zdonik SB (2003) Aurora: a new model and architecture for data stream management. VLDB J 12(2):120–139

Abadi DJ, Ahmad Y, Balazinska M, Çetintemel U, Cherniack M, Hwang JH, Lindner W, Maskey A, Rasin A, Ryvkina E, Tatbul N, Xing Y, Zdonik SB (2005) The design of the Borealis stream processing engine. In: Proceedings of the CIDR, Asilomar, pp 277–289

Aberer K, Hauswirth M, Salehi A (2006) A middleware for fast and flexible sensor network deployment. In: Proceedings of the VLDB'06, Seoul, pp 1199–1202

Ali MH, Chandramouli B, Raman BS, Katibah E (2010) Spatio-temporal stream processing in Microsoft StreamInsight. IEEE Data Eng Bull 33(2):69–74

Arasu A, Cherniack M, Galvez E, Maier D, Maskey A, Ryvkina E, Stonebraker M, Tibbetts R (2004) Linear road: a stream data management benchmark. In: Nascimento MA, Özsu MT, Kossmann D, Miller RJ, Blakeley JA, Schiefer KB (eds) Proceedings of the 30th international conference on very large data bases (VLDB), Toronto. Morgan Kaufmann, pp 480–491

BASt (1999) Merkblatt für die Ausstattung von Verkehrsrechnerzentralen und Unterzentralen (MARZ). Bundesanstalt für Straßenwesen. (in German)

Biem A, Bouillet E, Feng H, Ranganathan A, Riabov A, Verscheure O, Koutsopoulos HN, Moran C (2010) IBM InfoSphere streams for scalable, real-time, intelligent transportation services. In: Elmagarmid AK, Agrawal D (eds) Proceedings of the ACM international conference on management of data (SIGMOD), Indianapolis. ACM, pp 1093–1104

Bifet A, Kirkby R (2009) Data stream mining – a practical approach. University of Waikato. http://www.cs.waikato.ac.nz/~abifet/MOA/StreamMining.pdf

Bifet A, Holmes G, Kirkby R, Pfahringer B (2010) MOA: massive online analysis. J Mach Learn Res 11:1601–1604

Bogenberger K, Belzner H, Kates R (2003) Ein hybrides Modell basierend auf einem Neuronalen Netz und einem ARIMA-Zeitreihenmodell zur Prognose lokaler Verkehrskenngrößen. Straßenverkehrstechnik 47(1):5–12. (in German)

Chen H, Grant-Muller S (2001) Use of sequential learning for short-term traffic flow forecasting. Transp Res Part C Emerg Technol 9(5):319–336

Domingos P, Hulten G (2000) Mining high-speed data streams. In: Proceedings of the 6th ACM SIGKDD international conference on knowledge discovery and data mining, Boston. ACM, pp 71–80

Fiege G, Gasser T, Gehlen G, Geisler S, Jodlauk G, Phan MA, Quix C, Rembarz R, Wiecker M, Westhoff D (2011) ITS services and communication architecture. Deliverable D03, Cooperative Cars eXtended

Fontaine MD, Smith BL (2007) Investigation of the performance of wireless location technology-based traffic monitoring systems. J Transp Eng 133:157–165

Geisler S, Quix C, Gehlen GG, Jodlauk G (2009) A quality- and priority-based traffic information fusion architecture. Proc. of the 16th World Congress on intelligent transport systems and services (ITS), Stockholm, Sweden

Geisler S, Chen Y, Quix C, Gehlen G (2010) Accuracy assessment for traffic information derived from floating phone data. Proc. of the 16th World Congress on intelligent transport systems and services, Busan, Korea

Geisler S, Weber S, Quix C (2011) An ontology-based data quality framework for data stream applications. In: Proceedings of the ICIQ, Adelaide

Geisler S, Quix C, Schiffer S, Jarke M (2012) An evaluation framework for traffic information systems based on data streams. Transp Res Part C 23:29–55

Kargupta H, Sarkar K, Gilligan M (2010) MineFleet®: an overview of a widely adopted distributed vehicle performance data mining system. In: Proceedings of the 16th ACM SIGKDD international conference on knowledge discovery and data mining, Washington. ACM, pp 37–46

Khan A (2007) Intelligent Infrastructure-based queue-end warning system for avoiding rear impacts. IET Intell Transp Syst 1:138–143

Klein A, Lehner W (2009) Representing data quality in sensor data streaming environments. ACM J Data Inf Qual 1(2):1–28

Oza N (2005) Online bagging and boosting. Proceedings of the IEEE International conference on systems, man and cybernetics, Waikoloa, Hawaii, USA, pp 2340–2345. IEEE. http://dx.doi.org/10.1109/ICSMC.2005.1571498

Schmidt S, Berthold H, Lehner W (2004) QStream: deterministic querying of data streams. In: Nascimento MA, Özsu MT, Kossmann D, Miller RJ, Blakeley JA, Schiefer KB (eds) Proceedings of the thirtieth International conference on very large data bases (VLDB), Toronto. Morgan Kaufmann, San Francisco, Toronto, Canada, pp 1365–1368. http://www.vldb.org/conf/2004/DEMP29.PDF

Statistisches Bundesamt (2011) Unfallentwicklung auf deutschen Straßen 2010. http://www.destatis.de

Geospatial Visual Analytics of Traffic and Weather Data for Better Winter Road Management

Yuzuru Tanaka, Jonas Sjöbergh, Pavel Moiseets, Micke Kuwahara, Hajime Imura, and Tetsuya Yoshida

Abstract Sapporo is a city with two million citizens that gets 6 m of snow per year. This means that winter road management is very important for sustaining economic and social activities during the winter. We believe that an exploratory and iterative analysis and visualization approach is useful to support the decision making, to improve the winter road management strategies. We propose using a huge library of tools and services, and a framework that allows users to freely federate tools and services improvisationally ("mash-up") to create custom visualization and analysis environments and to apply these on appropriately selected data sets. Unlike conventional macro analysis approaches, we focus on micro analysis of winter road conditions. We use probe car data, speed readings etc., automatically collected from taxis and private cars. Geospatial visualization of the average speeds of all the road segments shows how different roads are affected by heavy snowfall, by snow plowing, and by snow removal. Combining geospatial visualization with knowledge discovery algorithms is a potential approach in this area. An example would be clustering the road segments based on similarity of the impact snowfall has to group roads into groups that can be maintained using similar strategies.

Keywords Exploratory visual analytics • Geospatial data visualization • Data mash-up • Improvisational federation • Winter road management

Y. Tanaka (✉) • J. Sjöbergh • P. Moiseets • M. Kuwahara • H. Imura • T. Yoshida
Hokkaido University, Sapporo, Hokkaido Prefecture 060-0810, Japan
e-mail: tanaka@meme.hokudai.ac.jp; js@meme.hokudai.ac.jp; moiseets@meme.hokudai.ac.jp; mkuwahara@meme.hokudai.ac.jp; hajime@meme.hokudai.ac.jp; yoshida@meme.hokudai.ac.jp

G. Cervone et al. (eds.), *Data Mining for Geoinformatics: Methods and Applications*, DOI 10.1007/978-1-4614-7669-6_6, © Springer Science+Business Media New York 2014

105

1   Introduction

1.1   Winter Road Management in Sapporo

Snow plowing and snow removal are fundamental public services for sustaining economic and social activities during the winter in Sapporo. Sapporo is the fifth largest city in Japan and has a population of 1.9 million people, and an annual cumulative snowfall of about 6 m with the maximum depth of snow cover reaching about 1 m. The local government's annual budget for snow plowing and removal is about 15 billion yen. Sapporo is one of few metropolitan cities with this severe snowfall. Others cities with heavy snowfall include Saint Petersburg, Harbin, Montreal, Ottawa, Helsinki, Calgary, Toronto, Syracuse, Anchorage, Buffalo, Rochester, Denver, etc., though none of them come close to Sapporo considering both population and snowfall.

Car traffic is seriously influenced not only by heavy snowfall but also by huge piles of snow on both sides of the roads. These piles (or *windrows*) of snow are created by the snow plowing. In Sapporo, both daily activities of residents and business and industrial activities depend heavily on automobile mobility.

Inadequate winter road management significantly slows traffic flow because of icy roads and roads being narrowed by windrows, which leads to further traffic congestion. Congested traffic will compress the snow and turn it into ice, making the roads extremely slippery; common around pedestrian crossings where cars have to break and accelerate. Repeated melting and freezing of roads by heavily congested traffic will also make the icy roads bumpy. This makes it difficult to steer, and also causes damage to the underside of the cars when it scrapes against the ice.

Studded tires were widely used in Japan, but the use was regulated in early 1990s to solve air pollution problems caused by the dust stirred up by studded tires. As a result, air quality improved significantly. However, it also resulted in several traffic issues in winter; e.g. an increase in car accidents on icy roads, slower traffic flow, increased use of anti-freezing agents and abrasives, and a significant increase of road management costs. Asano et al. estimated the direct and indirect economic losses from the ban on studded tires (Asano et al. 2002). The estimated annual losses were more than 18.5 billion yen in the Sapporo area. The main causes were the increase in driving times and costs, road accidents, and costs for maintenance and management.

1.2   Quantitative Studies of Winter Traffic

Takahashi et al. pointed out that in order to study traffic issues in winter it is important to understand the traffic patterns in a quantitative manner (Takahashi et al. 2004). The traffic patterns during weekdays notably differ from those during weekends or holidays. The traffic in winter also seems to vary greatly depending on specific weather conditions, e.g. snow cover, snowfall, and declining temperatures.

Takahashi et al. used taxi probe car data (taxi GPS data) as an advanced survey method to analyze traffic. They used multiple linear regression (MLR) analysis to explain the reduction in average travel speed (ATS) in winter compared to the ATS in summer in terms of explanatory variables including snow depth, snowfall amount, snowfall amount on the previous day, average temperature, and sunshine duration. Their MLR analysis showed that decrease in average temperature has the largest effect on reduced average travel speed in winter. Their counterintuitive result is that the amount of snowfall during the day or the previous day has little effect on the ATS. This conclusion however depends on their macro analysis, using only the average speed of the whole city.

Munehiro et al. from the same laboratory estimated the time loss along a specific route due to traffic winter congestion, using taxi probe car data (Munehiro et al. 2012). They also estimated benefits of snow removal on the same route. The analysis is still macroscopic.

The number of taxis and private cars from which probe car data can be obtained in real time has increased a lot in Japan. Probe car data are becoming fundamental in quantitative micro analysis of traffic changes caused by snowfall and icy roads.

Probe car data has been used for various purposes. One example of the use of probe car data for disaster management occurred after the Tohoku earthquake and tsunami in Japan in 2011. Car manufacturers in Japan have car navigation systems in the cars as standard, and these can use the driver's cell phone or other means of wireless communication to report back the car location and speed. These data are commonly used to provide services such as information on where there are traffic jams. Many roads were destroyed or blocked by the tsunami and earthquake. Three large car manufacturers (Honda, Toyota, and Nissan) provided the government with their probe car data. This could then be used to determine which roads were still being used, thus not destroyed or blocked, and which were not used, thus needing cleanup or restoration.[1] This information was also useful for planning where supplies could be sent, and other routing.

For other cities with severe snow conditions during winter, there have been scientific and technological research studies using ICT and GIS approaches to improve winter road management since the mid 1990s (Perrier et al. 2006a,b, 2007a,b). The objectives have been to provide heuristic-optimization solutions for many complex planning decisions. The main strategic and operational problems include defining a service level policy, locating depots, designing sectors, routing service vehicles, configuring the vehicle fleet, and scheduling the use of the vehicles. These activities are interrelated, and the effect that each decision has on the other decisions impacts the ability to provide the desired level of service. Since, unlike in Sapporo, probe car data from taxi cars or private cars (including both retrospective and real time data) are not yet available on a large scale, the actual application of the proposed methods in the real world has not been quantitatively well evaluated by e.g. calculating the change in traffic for different decisions.

---

[1]http://www.its-jp.org/saigai/ (in Japanese), accessed 2012-04-10.

Several measures to compare the road maintenance (snow removal etc.) performance was presented in Thill and Sun (2009) and used for comparing performance at two different highway segments in Buffalo, NY, during and after snow storms. These measure use Automatic Vehicle Identification (AVI) technology to collect vehicle speeds, and then use e.g. the relative decrease in speed or the time until the average speed has gone back to normal as performance measures. The paper also gives a good overview of previous research on the impact of snow on traffic conditions, and of research on performance measurements for snow and ice control.

1.3   Our Approach to the Problem

In this paper we use retrospective probe car data from taxis and private cars, and combine them with other data sources such as meteorological sensor data, snow plowing and removal records, complaints to call centers, and social media data from Sapporo. We propose a geospatial visual analytics environment for micro analysis of the relationships between different data sources by integrating geospatial visualization with data mining and clustering algorithms. Our goal is to achieve more efficient winter road management, better service or lower costs, with the help of information technology.

Unlike the conclusion of the macroscopic analysis by Takahashi et al. (2004), our microscopic analysis and visualization of the average speed reduction in each road segment indicates that it heavily depends on the amount of snowfall, the snow plowing, and the following snow removal. Heavy snowfall is immediately followed by snow plowing, which may result in huge piles of snow on both sides of the roads, making them narrow, which in turn worsens the traffic flow. Snow removal may take more than 1 day, but gradually improves and finally restores the traffic flow.

We provide geospatial visualization of the change of the average speed in each road segment, providing a visual environment for micro analysis of how roads are influenced by snowfall, snow plowing, etc. We also apply clustering algorithms to the change of the average speed in each road segment during snowy days (or weeks) to find groupings of road segments based on similarity of how the traffic is influenced by heavy snow. Such a clustering results may give insights into which roads can be managed in similar ways, and other strategic planning decisions.

Black ice areas around intersections are very slippery and dangerous, both for cars and for crossing pedestrians. Some automobile companies include the activation record of the ABS (Anti Lock Brake System) in their probe car data, which can then be used to detect slippery intersection areas. The probe car data we have do not include the ABS activation records. However, a comparison of the speed distribution histograms between winter and summer may tell us which intersections are slippery. Slippery intersections can be expected to have a large peak in speeds from 0 to 5 km/h, while non-slippery intersections would have histograms similar to the histograms for summer.

We believe that the problem of optimized or better snow plowing and removal is not a simple system that can be modeled by a single monolithic mathematical model. It is a complex system composed of many mutually interrelated systems. Therefore, any macroscopic analysis that applies statistical analyses or knowledge discovery algorithms such as data mining to the whole system may miss significant insights. Statistical analyses and knowledge discovery algorithms assume application to well-formed problems that can be mathematically modeled. One reason why we need micro analysis is that the problem at hand is not well-formed nor easy to model.

We believe that we first need to iteratively refine our analyses depending on observations and evaluations of previous analysis results. Trial-and-error exploration of the data using visualization and analysis tools can help us get a better understanding of the problem and what data to analyze, which questions to ask, etc. Once we have gotten this far, the problem can be considered to consist of subproblems, each of which is approximately well formed, and normal statistical analysis tools or knowledge discovery algorithms can then be applied to them.

We propose an integrated geospatial visualization and analysis environment for such exploratory visual analytics. Such a system requires a large library of visualization and analysis tools. It should support improvisational composition ("mash-up") of whatever visualization or analysis environments may be required for a specific problem, i.e. it should allow free recombination of tools etc. Our integrated environment uses the Webble World as its enabling technology. Webble World is a Web-top system version of the Meme Media architecture which was first proposed in 1993 and has been extensively studied since then. Webble World uses visual components called Webbles that can be improvisationally federated together by users to compose complex applications without the need for coding.

In order to bring GIS, statistical analysis tools, knowledge discovery tools, and SNS (Social Networking Services) like Twitter into the Webble World and to make them interoperable, we created a wrapper for ArcView, a generic wrapper for tools written in R and Octave, and a generic wrapper for Web services. This wrapping allows us to improvisationally federate for instance the Twitter service with geospatial visualization of probe car data within no more than a few minutes. The exploratory visual analytics with improvisational federation of visualization and analysis tools in a large library will provide a new integrated environment for micro analysis of how traffic is influenced by snowfall, snow plowing, and snow removal.

2   Enabling Technology

2.1   Webble World

We create our system using a technology called Webbles. Webbles are software objects intended to make sharing and re-use of functionality and services as easy as copying and pasting texts and images is today. Webbles are the latest incarnation of

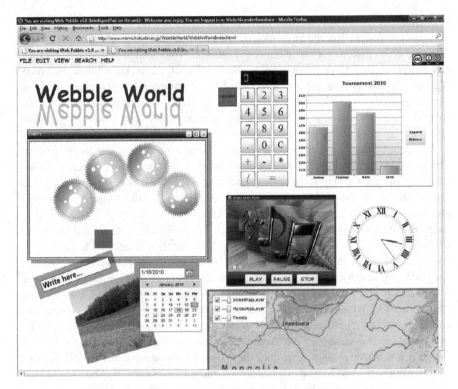

Fig. 1 A few Webbles loaded into a browser in order to display their variety and versatile usefulness. Each single piece or object is a standalone Webble entity; a building block that can be used to interact and collaborate with other Webbles

the IntelligentPad idea. The original idea is based on Richard Dawkins's concept of the Meme (Dawkins 1976), the idea that thoughts, knowledge, and ideas reproduce and mutate in ways similar to what happens to biological genes, and that recombination of ideas can lead to new and perhaps even more useful ideas.

Based on this concept, the Meme Media and IntelligentPad concepts were born (Tanaka 2003). An IntelligentPad is a digital object that can exist in various forms and have multiple purposes, but always has a standardized interaction interface. Several generations of IntelligentPad frameworks have been developed and experimented with during the last decades, but here we focus on the latest version, the Webble World (Fig. 1).

A Webble (Kuwahara and Tanaka 2010) is a customizable object, and Webbles can be loaded from online Webble repositories. Webbles run inside Web browsers (any browser that supports Microsoft Silverlight). Webbles communicate primarily through "slots". A slot is a wrapper that works as the interface for a property of an object or a method it supports. A simple example could be a text label Webble, that can have a slot for the text to display, another slot for the font to use, another for the background color etc.

Slots are named slots because you can plug Webbles together using slots. When you connect two slots the properties of these are synchronized, so when the slot value changes in one object the other object is automatically updated too. A simple example of this could be connecting the "title" slot of a music playing Webble to the "text" slot of a text label to display the title of the song that is being played. When the music playing Webble starts playing another song, the label text will automatically change. For more advanced objects, more interesting two-way communication can also be useful.

Webbles are structured in parent-child hierarchies, which controls the communication structure. This makes it easier to analyze the information flow in complex applications built using Webbles.

There are many Webbles available for use already. Examples include simple components like text boxes, drop down lists, labels, etc.; more complex Webbles include interactive maps, charts, intelligent windows, movie players, etc. Some Webbles allow increased control of other Webbles or the surrounding browser environment, and others allow access to online databases, or allow connecting to Web services.

Freshly created Webbles are called primitive Webbles. Webbles can also be combined together to form more complex objects. Such objects can be saved as a package and are then called compound Webbles. Saving a set of primitive and compound Webbles that are not all connected together is also possible, and this is called a Webble application. Primitive Webbles, compound Webbles, and Webble applications can then be loaded and used by anyone, and it is possible to combine them with Webbles from other sources, or to pick apart complex Webbles and re-use only some parts for new purposes. The idea is that these pluggable components can be used like a "Meme Lego", with simple building blocks that can be recombined to build many different things (Fig. 2).

Primitive Webbles that need to be developed to introduce some functionality not already available in the Webble world are built by programming in Microsoft Silverlight (a subset of Microsoft C#). There are templates available that build the general Webble interface, so only the parts specific to the new functionality need to be written by the programmer. Primitive Webbles like this can for instance be used to build a Webble interface to wrap already existing software. This has been done for instance for the R language for statistical analysis. So you can run R programs inside Webble world, and connect them to the other available Webble components.

Compound Webbles and Webble applications are intended to be easy to create without the need for programming knowledge. The simple interface with slots that you can connect while running in the browser allows all types of users to plug together existing components to solve new problems.

The Webble World is available for use by anyone online.[2]

---

[2]http://www.meme.hokudai.ac.jp/WebbleWorld/WebbleWorldIndex.html
Username: webble, Password: webble.

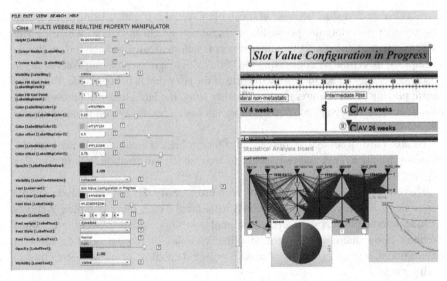

Fig. 2 Just by altering slots values and setting up connection paths between specific Webbles you may build full-fledged rich Internet applications directly online and share them with the world

2.2 Mapping Technology

When dealing with information that is location dependent, it is often useful to display the information on a map, e.g. showing which roads seem to have traffic problems by drawing those roads in red in an overlay over a map of the city. While the same information could be given in other ways, e.g. a table with street names of problem areas, a map is a familiar way to get an overview of the situation quickly.

There are several mapping software toolkits available. One example is the Google Earth (and Google Maps) API[3] from Google. Using a standardized XML format called KML, it is possible to specify for example points, lines, and polygons to be displayed in 3D on a satellite image of the earth. You can also freely position the viewpoint to specify the part of the earth to display, from what angle, the zoom, etc. It is possible to link images, videos, and other data to points on the map too.

Another mapping toolkit is the ArcGIS suite from the Environmental Systems Research Institute's (ESRI). We have wrapped the ArcGIS API for Silverlight[4] into our Webble World framework (Sect. 2.1), and it is thus possible to combine

---

[3]Google. (2012) Google Earth (Version 6.2.1.6014 (beta))
http://www.google.com/earth/download/ge/agree.html (Accessed 6 April 2012).

[4]ESRI. (2012) ArcGIS API for Silverlight (Version 2.4)
http://help.arcgis.com/en/webapi/silverlight/

a mapping tool Webble based on ArcGIS with all the other available software components (Webbles). The Webble wrapper for the ArcGIS has slots for specifying various mapping layers, size of the map, what sections of layers to display, etc.

3   Clustering Taxi Probe Car Data

3.1   Background

Dataset

The taxi probe car dataset was kindly provided to us by Fujitsu Co. LTD. The dataset is based on the information provided by about 2,000 taxi cars, running in the urban area of Sapporo. Roads are split into road segments, with a new road segment starting at each intersection. Statistics are provided for each road segment every 5 min. These include: the average speed, the top speed, the number of cars that passed, the length of the segment, etc. There are around 120,000 road segments in the provided data. We have data for two periods: the *snowfall period*, from January 1st, 2011 to February 7, 2011; and the *non-snowfall period*, from September 19, 2010 to September 25, 2010.

We investigated the influence of snowfall on the traffic in the dataset. Even when the average speed of a road segment is high in the non-snowfall period, the average speed during the snowfall period might be much lower when the road is covered with snow and ice, or narrowed by piles of snow on the sides. Since the dataset characterizes each road segment in terms of its statistics, road segments can be divided into several clusters based on similarities of these statistics. These groups might then give insights on e.g. roads that are suitable for similar snow removal strategies.

Preprocessing

As detailed above, each road segment is represented by the statistics for the segment gathered every 5 min. In order to get a representation for the segment for 1 day, we simply concatenated the average speed statistics (though other statistics could also be used) of each 5 min period into one high-dimensional vector. This vector does not contain information on the physical properties of the segment, such as the width of the road, the number of lanes, or which other segments are adjacent; it only contains the average speed data. When clustering on the average speed, the a vector would contain 288 speed readings, on for each 5 min period during the day.

In addition, since the probe car data is transmitted via a radio system from the moving taxis in the real world, the dataset suffers from lots of missing values. In order to calculate the similarities among vectors, it is necessary to fill in the missing values. How these missing values are assigned affects the results of analysis. As a preliminary experiment, we filled in the missing values of any road segment with the average value of that segment.

Clustering Method

Since each road segment is represented by a high-dimensional vector, we utilized the spherical k-means algorithm (skmeans) (Dhillon and Modha 2001) to clustering the road segments. This algorithm is an extension of the standard k-means algorithm (Hartigan and Wong 1979), and was proposed for the clustering of high-dimensional data such as text. In the standard k-means algorithm, each vector is assigned to the "nearest" cluster in terms of Euclidian distance. However, when the number of dimensions get large, the performance of k-means degrades due to the "curse of dimensionality". In order to remedy the performance degradation, cosine similarity, which is invariant to the number of dimensions, is utilized as a similarity measure in the skmeans algorithm. Thus, each vector is assigned to the "nearest" cluster in terms of cosine similarity.

The true number of clusters is unknown in the dataset. Thus, we varied the number of clusters from 4 to 10 in the following experiments. The results are shown when clustering the data into 6 clusters.

The clustering is only done on speed data. There is no following clustering using the spatial (geographical) information, and the vectors do not contain any such information. This means that the clusters are not (necessarily) spatially connected, and that they can overlap in any way in the physical world.

3.2 Results

The Difference Between Snowfall and Non-snowfall Periods

In order to see if there is any difference between the snowfall and the non-snowfall periods, we created a vector in 288 dimensions for each road segment by concatenating the average speeds of the segment during 1 day (60 min × 24 h / 5 min = 288). A set of such vectors for the road segments is constructed for each day, and clustering with skmeans was conducted for each set of vectors.

Figure 3 shows the clustering results of a day in the non-snowfall period (the number of clusters was set to 6). Normally, the clusters are given different colors and can be overlaid, but here we show the results in grayscale, and show one cluster per image frame. (Figure 7 shows clusters overlaid on top of each other, with one cluster shown in bold for contrast.)

As an example cluster, the second cluster in the top row in Fig. 3, which has all lines shown in **bold**, correspond to the road segments that were clustered together. The depicted lines correspond fairly well with some of the main (arterial) roads

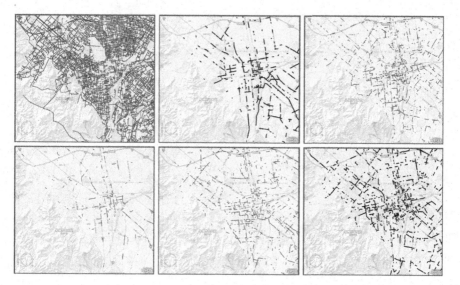

Fig. 3 Clustering result in the non-snowfall period, one cluster per image. Lines in **bold** show clusters with road segments mainly from the main (larger) roads of the city

in Sapporo where many cars run at rather high speed. Note that even though physical connectivity relations among road segments are not utilized in the vector representation, many of the road segments clustered together connect nicely into lines with connected road segments, based solely on their average speed statistics. The last cluster, also shown with bold lines, also contains many road segments from the arterial roads. These two clusters together cover almost all the segments of the arterial roads.

Figure 4 shows the clustering results of a day when heavy snowfall occurred (again, the number of clusters was set to 6). Compared with Fig. 3 we can see that the lines for the clusters with **bold** lines, i.e. the clusters covering the major roads, are more fragmented. Segments from the same road now end up in different clusters more often. This could be caused by traffic jams occurring in some parts of the road segments due to the snowfall, especially in segments where the road has become more narrow than normal because of piles of snow occupying parts of the road after snow plowing. In the next section, we take a closer look at this.

The Difference Before and After the Snow-Removal

The results in Figs. 3 and 4 indicate that the profiles of a road segment are different in snowfall and non-snowfall periods. We believe that this difference in large parts is caused by traffic jams caused by the snowfall. Since piles of snow narrowing parts

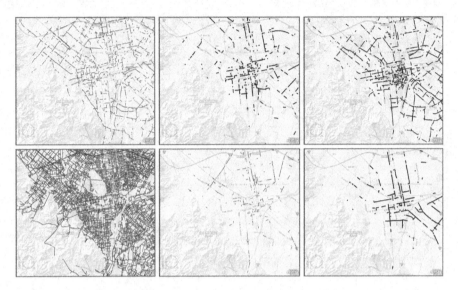

Fig. 4 Clustering result for the snow period, one cluster per image. Lines in **bold** show clusters with road segments mainly from the main (larger) roads of the city

of the roads is mostly a problem during rush hour (when few cars are on the road, the road being narrow is not a serious problem), in the following experiment, we focused on data from 7:00 to 10:00 a.m.

In order to highlight the difference between snowfall and non-snowfall periods, we concatenated the average speeds during the rush hour period of 1 day in the non-snowfall period with the speeds during rush hour for 1 day during the snowfall period, thus creating another type of vector of 72 dimensions (60 min × (3 h + 3 h)/5 min = 72). We created such vectors for two days during the snow period: the day directly after a heavy snowfall, and the day two days after the same snowfall.

Figure 5 shows the results for the day just after the heavy snowfall occurred, and Fig. 6 shows the clustering result of the next day (and no snowfall in between). A closer look at an example road is shown in Fig. 7. The **bold** line running vertically is one of the main roads through the city, and right after the heavy snowfall, segments from this road end up in four different clusters. The next day, after snow removal, all the segments in the road are clustered together again.

By comparing these results, we can see that the day after the heavy snowfall had the main roads fragmented, with segments ending up in many different clusters, but the fragmentation is reduced after snow removal. This confirms our belief that snowfall leads to traffic problems and that when the snow is removed, the problems also go away. The results also indicate the importance of microanalysis of traffic to discover the best strategies for road management and maintenance specific to different types of roads.

Fig. 5 Clustering result for rush hour of a day directly after a heavy snowfall. The segments from the major roads end up in several different clusters

Fig. 6 Clustering result for rush hour after snow removal (2 days after the snowfall). The segments from the major roads are generally clustered together

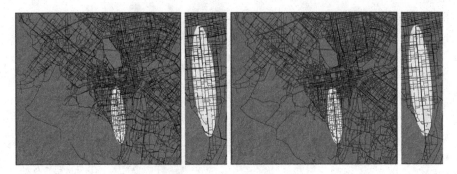

Fig. 7 A closer view of the clustering results of Figs. 5 and 6

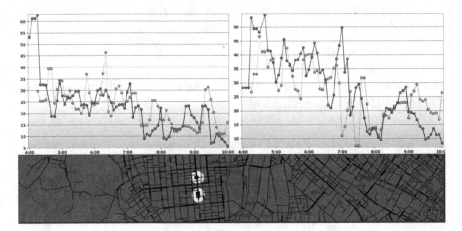

Fig. 8 Two road segments that are clustered together during snow free days but end up in different clusters after snowfall are highlighted. Time series of average speeds for the two road segments are shown in two different graphs. The *left half* are readings from the non-snowfall period, and the *right half* are from the snowfall period. Each graph has one *curve* for the day with heavy snowfall (*light gray*) and one for the snow free day after that (*dark gray*)

3.3 Combining the Cluster Results with Other Visualization Methods

In Sect. 3.2, we saw that road segments belonging to the same major road were clustered into the same cluster during the non-snow period and during snow free days of the snowy period. On days with heavy snowfall, and the snowy period overall, segments from the same road ended up in several different clusters, though.

Our system makes it possible to improvisationally federate additional data sources and data visualization tools, to take a closer look at the data or visualize it in other ways. One example of this is shown in Fig. 8. Two road segments that were clustered together during the snow free day in Fig. 7 (and belong to the same road) are highlighted and drawn thicker than the other road segments. These two segments were clustered into different clusters on the day after the heavy snowfall (Fig. 7).

It is possible to bring up graphs showing the average speed of the road segments as a function of the time of day. The speed during different times of the day is what the clustering is based on. There is one graph for each road segment, and each graph shows two curves. The dark gray curve shows the speeds during the snow free day, and the light gray curve shows the speed during the day with heavy snowfall. We can see that the curves for the snow free day are at least fairly similar in the right half of the graph (the left half are the concatenated data from the non-snowfall period, the right half is from the snowy period), while the light gray curves are very different, as can be expected based on the clustering results in the previous section.

4   Histograms of Speeds in Slippery Intersections

Intersections are especially prone to bad road conditions during the winter in Sapporo. Cars have to break and stop when the light is red, and they have to accelerate and often slip, causing the tires to spin and polish the compressed snow and ice into very slippery ice. The intersections also usually become very uneven and bumpy, making it even more difficult to drive safely and without damaging the car.

Not all intersections are affected the same way, and knowing where there are problems could help prioritize where to go to put more sand (to reduce slipperiness) or file down icy bumps. Here we show speed histograms for cars when they are driving through various intersections around our university. The speed readings are taken from private cars equipped with car navigation systems (collecting the speed readings) and cell phones with subscriptions to a service notifying the navigations system of possible traffic jams etc.

In Fig. 9, speeds are shown as speed histograms, one, in black, pointing up, for the speed in the summer (good road conditions) and one, in gray, pointing down, for the speed in the winter (possibly slippery conditions). The leftmost bar in each histogram is the relative amount of cars driving 0–5 km/h, the second bar from the left is for 5–10 km/h, and so on. The histograms are scaled so that the volume of each histogram is the same, so the bars do not show the number of cars driving slowly but the proportion of cars that traveled slowly.

The rightmost intersection of the two top intersections with histograms is famous for being slippery. We can see that the gray histogram has a larger proportion of cars traveling at slow speeds. In contrast, the leftmost intersection in the lower part of the picture rarely becomes slippery, and the gray histogram there does not have a high proportion of cars at slow speeds.

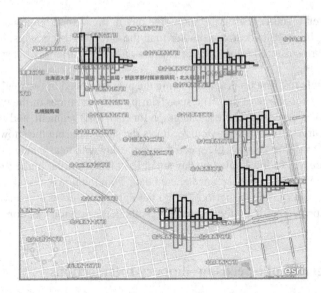

Fig. 9 Histograms showing the speed in intersections around our university. The *gray bars* are the speed in winter and the *black bars* are the speed in summer. The *bar* farthest to the *left* is 0–5 km/h, the next 5–10 km/h, etc

5 Mashups with Probe Car Data and Twitter

For disaster management, combining many different types of data sources can be useful. We do not necessarily know in advance exactly what data will be of use to us, and the data we want the most may not be available so we may have to make do with combining other sources to achieve the same goal. Here we show an example of combining and visualizing data from many different and diverse data sources.

Figure 10 shows a screenshot when visualizing several types of data. Here the visualization has been changed to colors showing up better in grayscale, but in actual use all data sources have clearly distinguishable colors instead. The following types of data are shown:

Probe car data. The streets are colored in white, gray, and black based on probe car data. Where the average speed is roughly the same as the speed in the summer, black is used. On streets where the speed is lower than in the summer gray is used, and when the speed is much worse than in summer the road is colored white. The data is from around 2,000 taxi cars reporting speed readings at each intersection they pass. The day shown in Fig. 10 had fairly bad snow conditions, so the traffic situation was not good and most roads are white or gray.

Weather station data. Snowfall measured at about 50 weather stations located in different parts of the city is shown as black bars standing alone, on top of the location of the weather station. Tall bars indicate large amounts of snow, low bars indicate little snow fall. The weather sensors also collect many other types of data not visualized here.

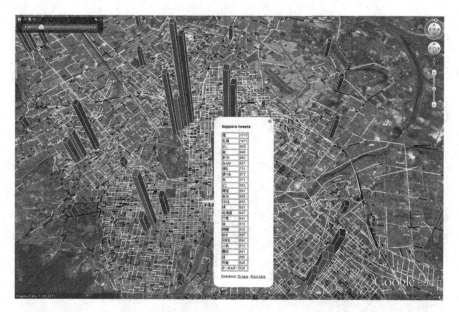

Fig. 10 A mash-up of many different data sources. The images show the probe car speed data, the most common words from Twitter tweets in Sapporo, weather data, snow removal data, and snow removal complaints to call centers. This visualization was done using Google Earth, not using our Webble World framework

Call center complaints. The amount of complaints from citizens calling the local snow removal call centers or the city call centers are shown as two dark gray bars (a tall bar meaning many complaints) in a group of four bars. The complaints are the two rightmost bars, with the city call centers on the left, and the local snow removal call centers on the right. The data is shown for each ward (district) in the city, and the bars are located in a central location of the ward, usually the location of the ward office.

Snow removal data. Data from the snow removal companies is shown as two gray bars, also in the set of four bars placed in the central ward location. The leftmost bar indicates the road distance plowed by snow plowing vehicles. The second bar from the left indicates the amount of snow removed by trucks taking snow out of the city to specified dumping locations.

Twitter data. The most frequent words mentioned in Twitter tweets that are location tagged as coming from Sapporo are shown in a table. Here there is one global table for the whole city, but displaying local tables for tweets from smaller areas is also possible.

In actual use, showing all the information at once, as in Fig. 10, is probably not very useful since there is too much information being displayed at once. Thus, turning individual data sources on and off has been made easy. This data can then show for instance how the traffic situation was affected by the weather and the snow removal efforts.

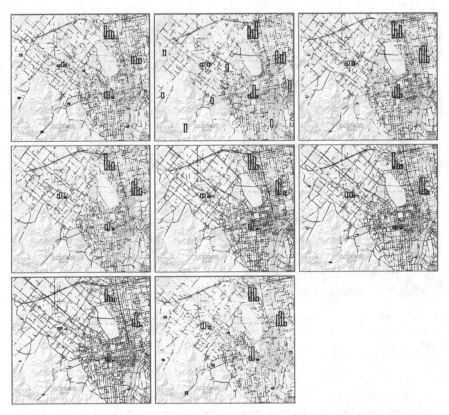

Fig. 11 A time series showing the change in the traffic situation after a heavy snowfall. Once the heavy snow hits, the traffic situation becomes bad and there is a large spike in complaints to call centers. The situation then gradually improves, and complaints go down. Then, more snow falls and the traffic situation worsens again. *White lines* indicate streets where the average speed is much worse than in the summer, and *black lines* are streets where the speed roughly the same as in the summer. There is one image per day, *left to right*, *top to bottom*, with the *top left* being January 31, 2011, and *bottom right* February 7

Figure 10 shows the data displayed using Google Earth. The same data can also be shown using the ArcGIS toolkit, which we have wrapped so it can be used in our Webble World framework. The following example scenario is shown using our Webble World.

The example scenario is shown in Fig. 11. The figure shows the days from January 31 to February 7, 2011. On January 31 we can see that some roads are white or gray, so the traffic situation is not as good as in the summer, which is to be expected, but there are not so many complaints coming in. On February 1, there was a very heavy snowfall as can be seen by the many high black bars standing alone all over the city (snowfall measured at weather stations). The traffic situation turns bad, and large parts of the streets in the city are white or gray. There are also spikes in the call center complaints data (the two rightmost bars of the sets of four bars placed together).

The following days there is no or very light snowfall, but the traffic situation does not improve over night. It takes several days before most of the city streets are back to black. The complaints are also still quite high for several days, even though there is no new snowfall. This is of course because there is still a lot of snow left from the snowfall on February 1, as can be seen for instance by the fact that the snow removal companies keep removing very large amounts of snow the following days too (the leftmost bars in the sets of four bars).

The complaints gradually go down as more and more snow is removed, and the traffic situation also goes back more or less to normal. On February 6, the second to last picture, the situation is good. Most of the streets are black (good traffic conditions), and there are few complaints. On February 7 (last day shown), it snows again, and the traffic situation turns bad again.

Since other factors also affect the traffic conditions, adding other data sources can also be useful. Figure 10 also had data from Twitter. This can be useful when something unexpected happens. On the February 7, the Sapporo Snow Festival started. This is a huge event where over a million tourists come to the city, and several big streets around the festival area in the middle of the city are closed off. This of course has a large effect on the traffic flow in the city. When noticing that traffic flow around the area is bad despite snow conditions being good and other streets not having problems, we could bring up a list of common words when people write status updates in Sapporo (or a smaller area). February 7 has "snow" and "festival" among the top five words, and we could guess that this is what is causing the problems.

While the Snow Festival occurs every year and is thus not an unexpected event, a similar approach can be used when unexpected conditions occur. Something that affects a lot of people is likely to be mentioned by many of them. Now that most people are constantly connected through their cell phones, information such as the status tweets on Twitter can be helpful. Our system allows combining data from such diverse sources as cell phone tweets, weather stations, and probe cars.

The strength is that as long as a data source, analysis method, or display method has been wrapped to work in the Webble World environment, connecting them in new ways is very quick and easy. If the Twitter information does not tell us enough, we can plug in a news feed subscription Webble and display words mentioned in news stories about Sapporo instead, with only a few minutes work. Or some other completely different information source.

6   Tweets

In Sect. 5, data collected from citizens writing status updates on Twitter was visualized together with many different types of data. There, we showed the aggregated statistics of all tweets from a specific area, displaying the most frequently mentioned words to see what seems to be going on in that area.

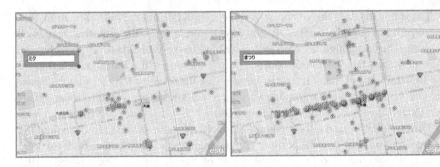

Fig. 12 Location tagged tweets

Instead of showing which words occur in an area, it is also possible to show all the locations of tweets mentioning a specific word. This can tell us where people are when they talk about different things. Not all tweets are location tagged, but many are (actually a small minority of all tweets, but still a fairly large amount in raw number of tweets). Some are tagged only with the general location, e.g. "Sapporo Station", and some have GPS coordinates.

Figure 12 shows two examples of displaying location tagged tweets. There is a text field to input a search string to indicate what tweets you want to visualize, and a map showing the locations of tweets containing this string.

On the left is a picture of the locations of tweets mentioning *"Miku"*. *Miku* is the name of a singing synthesizer persona, and there was a big snow sculpture portraying this pop singer character at the Sapporo Snow Festival. On February 7, this 3 m tall statue fell off its base and injured a tourist. Many tweets mentioning *Miku* can be seen around the West 5 area of the Odori park, which is were the statue was located. There are tweets mentioning the event from other locations too, of course, but there is a clustering effect around the area of the statue.

Similarly, on the right in Fig. 12 is a display of the tweets mentioning the Snow Festival. Most tweets are located in a band from west to east in the Odori park, which is the festival area.

Visualizing where people are mentioning that they are slipping, or that they fell, can tell us where the road conditions are bad. There are however not very many location tagged tweets mentioning such things. Of the 40,000–50,000 tweets tagged as coming from Sapporo each day, only around 800–900 have the specific location (GPS coordinates). Of these, only about one message per day mentions falling or slipping in our data. So slippery roads are not noteworthy or exceptional enough to most people and thus the Twitter feed cannot pick up on it, at least not yet (this might improve when more people start using their GPS enabled smart phones to send the messages). For more severe problems, it would likely be possible to notice them in the messages of ordinary people.

We believe using this type of information source can be very useful. The "end users" of the snow removal services are the citizens, so this information directly provided by the citizens themselves that can also automatically be collected is hopefully useful, and it has the potential to contain information not covered by the other information sources.

7 Conclusions

Winter road management and maintenance is fundamental in Sapporo for sustaining economic and social activities in winter. Since the influence of snow on traffic is a complex system that is hard to mathematically model as a single system, explorative and iterative analysis and visualization is necessary to support decision making for better management and maintenance strategies.

We proposed using a huge library of visualization and analysis tools and services, and a system framework supporting improvisational federation ("mash-up") of them to compose whatever analysis environment may be deemed necessary based on previous analysis. Unlike conventional macro analysis approaches, we focus on micro analysis of winter traffic conditions using taxi and private probe car data. We showed how visualization of changes in average speed of each road segment shows the influence from heavy snowfall, snow plowing, and snow removal. We also proposed a method to estimate which intersections may have problems with black ice from the probe car data.

We also applied a clustering method to probe car data to classify road segments based on similarity in the effect of heavy snow on the traffic of the segment. This could give insights into management decisions for different segments.

Acknowledgements This research has been supported by JST (Japanese Science and Technology Agency) in 2012 as a feasibility study on open smart federation of cyber physical data, and the application to smart snow plowing and removal.

We would like to thank Honda Motor Co., Ltd, and Fujitsu Ltd for providing us with sample probe car data. We also thank the Sapporo City Government for information and records on snow plowing and removal, and the Japanese Weather Association for meteorological data.

References

Asano M, Hara F, Tanabe S, Yokoyama S (2002) Evaluation of the ten years since studded tires were banned in Hokkaido and future issues, new challenges for winter road service. In: XIth international winter road congress, Sapporo, p 10
Dawkins R (1976) The selfish gene. Oxford University Press, New York
Dhillon JS, Modha DS (2001) Concept decompositions for large sparse text data using clustering. Mach Learn 42(1):143–175
Hartigan JA, Wong MA (1979) Algorithm AS 136: a k-means clustering algorithm. J Appl Stat 28(1):100–108
Kuwahara M, Tanaka Y (2010) Webble world – a Web-based knowledge federation framework for programmable and customizable Meme Media objects. In: The IET international conference on Frontier computing 2010, Taichung, pp 372–377

Munehiro K, Takahashi N, Watanabe M, Asano M (2012) Winter road traffic evaluation using taxi probe data in Sapporo, Japan. In: Proceedings of the 91st TRB annual meeting, Washington

Perrier N, Langevin A, Campbell JF (2006a) A survey of models and algorithms for winter road maintenance. Part I: System design for spreading and plowing. Comput Oper Res 33:209–238

Perrier N, Langevin A, Campbell JF (2006b) A survey of models and algorithms for winter road maintenance. Part II: System design for snow disposal. Comput Oper Res 33:239–262

Perrier N, Langevin A, Campbell JF (2007a) A survey of models and algorithms for winter road maintenance. Part III: Vehicle routing and depot location for spreading. Comput Oper Res 34(1):211–257

Perrier N, Langevin A, Campbell JF (2007b) A survey of models and algorithms for winter road maintenance. Part IV: Vehicle routing and fleet sizing for plowing and snow disposal. Comput Oper Res 34(1):258–294

Takahashi N, Miyamoto S, Asano M (2004) Using taxi GPS to gather high-quality traffic data for winter road management evaluation in Sapporo, Japan. In: Sixth international symposium on snow removal and ice control technology, Spokane, pp 455–469

Tanaka Y (2003) Meme Media and Meme Market architecture. IEEE Press, Piscataway

Thill JC, Sun H (2009) Comparing approaches to winter highway maintenance operations through user mobility performance. J Transp Res Forum 48(1):49–64

Exploratory Visualization of Collective Mobile Objects Data Using Temporal Granularity and Spatial Similarity

Tetsuo Kobayashi and Harvey Miller

Abstract Recent advances in location-aware technologies have produced vast amount of individual-based movement data, overwhelming the capacity of traditional spatial analytical methods. There are growing opportunities for discovering unexpected patterns, trends and relationships that are hidden in massive mobile objects data. However, a lingering challenge is extracting meaningful information from data on multiple mobile objects due to the visual complexity of these patterns even for a modest collection of mobile objects. This chapter describes visualization environments based on temporal granularity, and spatial and/or attribute similarity measures for exploring collective mobile objects data. Reconstructing trajectories at user-defined levels of temporal granularity allows exploration at different levels of movement generality. At a given level of generality, individual trajectories can be combined into synthetic summary trajectories or classified into groups based on locational and/or attribute similarity. A visualization environment based on the space-time cube concept exploits these functionalities to create a user-interactive toolkit for exploring mobile objects data. A case study using wild chicken movement data demonstrates the potential of the system to extract meaningful patterns from the otherwise difficult to comprehend collections of space-time trajectories.

Keywords Spatio-temporal knowledge discovery • Temporal granularity • Mobile objects • Data aggregation • Geovisualization

T. Kobayashi (✉)
Department of Geography, Florida State University, 600 W College Ave, Tallahassee, FL 32306, USA
e-mail: tkobayashi@fsu.edu

H. Miller
Department of Geography, University of Utah, 201 Presidents Cir, Salt Lake City, UT 84112, USA
e-mail: harvey.miller@geog.utah.edu

G. Cervone et al. (eds.), *Data Mining for Geoinformatics: Methods and Applications*, 127
DOI 10.1007/978-1-4614-7669-6_7, © Springer Science+Business Media New York 2014

1   Introduction

Location-aware technologies (LATs) such those based on the Global Positioning System (GPS) or radio-frequency identification (RFID) chips have greatly enhanced capabilities for collecting data about mobile objects. LATs connected to location-based services (LBS) embedded in cellular telephones and other clients allow unprecedented access to individual mobility patterns across a wide range of domains (Brimicombe and Li 2006; Li and Longley 2006). GPS and RFID devices are increasingly connected to vehicles and objects in fleet management and logistics, generating fine-grained data on movements of these entities within supply chains (see Roberti 2003). Researchers in ecology and biology are also using LATs to track movements of animals, creating new insights into territoriality and ecosystem dynamics (e.g., Wentz et al. 2003; Turchin 1998)

The prevalence of LATs is generating a vast amount of mobile objects data that are overwhelming the capabilities of traditional spatial analytical methods. A major challenge in GIScience is to develop representation and analysis techniques that can handle spatio-temporal and mobile objects data (Laube et al. 2005). A related research challenge is developing methods to explore, analyze and understand the motions of collections of mobile objects over time. Analyzing one or a small number of mobile objects is tractable, but making sense of the collective mobility patterns of even a modest number of objects is daunting due to the visual complexity of the data involved (Shaw et al. 2008).

This paper describes a user-interactive visualization toolkit for summarizing and exploring mobile objects data based on spatial similarity among object trajectories at different levels of temporal granularity. Reconstructing trajectories at user-defined levels of temporal granularity allows exploration of the mobile objects at different levels of movement generality. At a given level of granularity, the user can apply similarity measures for aggregating or grouping trajectories based on location or spatial similarity. To maximize user-interactivity, the measures are computationally scalable to facilitate rapid calculation even on modest computational platforms. The similarity measures are also dimensionless and semantically-clear to facilitate easy interpretation. A visualization toolkit based on the space-time cube concept exploits these functionalities to create a user-interactive environment for exploring mobile objects data. A case study using wild chicken movement data demonstrates the functionality of the toolkit for extracting general patterns from an otherwise indiscernible collection of trajectories.

The visualization environment described in this paper intends to provide a user-friendly toolkit for scientists who are primarily concerned with data corresponding to objects moving through geographic space such as people, vehicles and animals. Therefore, we designed the methods and toolkit in this paper for objects that exhibit potentially continuous motion through space densely with respect to time. The temporal granularity and similarity aggregation methods are not designed for objects that exhibit discontinuous change at discrete moments in time. Therefore, other event or change data such as financial transactions or phone calls, while often referenced in time and/or space, are not appropriate.

The next section of this paper provides the background behind the concepts and methods used in this research. Following this, the methodology section explains the temporal granularity, similarity functions, and other visual and summarization functionalities of the visualization toolkit. Section 4 provides a case study that illustrates the methods using a space-time visualization toolkit. The final section summarizes the contributions of this research and suggests topics for further investigation.

2 Background

2.1 GIS and Mobile Objects

Since change with respect to time is common in most natural and human phenomena, incorporating time and change in geographic information systems has been a critical research frontier since the 1980s (see Langran 1992). Possible approaches include temporal "snapshots" that update the database at regular intervals, event-based approaches that update only the relevant portion of the data when a change occurs, and maintaining semantic, spatial and temporal dimensions in separate but linked domains (see Peuquet and Duan 1995; Yuan 2001; Worboys and Duckham 2004; Hornsby and Cole 2007).

Mobile objects data is an important special case of this general problem since these change their geometry near-continuous with respect to time. Explicitly updating of the geometry of a moving object is too expensive with respect to computational effort and storage requirements. Instead, one must accept some level of sampling error due to the finite and discrete updating of continuously changing objects and represent this error within the database (Sistla et al. 1998). Mobile objects also imply unique semantics and therefore a need for specialized query languages and analytical techniques (Andrienko et al. 2008). The field of mobile objects databases has emerged to handle these unique requirements.

Another set of techniques for understanding mobility data derives from the field of time geography and efforts to build GIS and other analytical tools based on its basic concepts. Time geography is based on the notion that the events that comprise an individual's or object's existence have spatial and temporal dimensions that are difficult to untangle in a meaningful way (Hagerstrand 1970). While only a conceptual framework traditionally, in recent years the applicability of time geography has been enhanced through the development of analytical and computational tools linked to GIS software (e.g., Kwan 2000; Miller 1991, 2005; Yu and Shaw 2008). However, most of these efforts address only a single or small number of mobile object trajectories due to time geography's bias towards the individual rather than collective behavior, as well as a lack of tools for handling collections of trajectories. For example, although Kwan (2000) develops interactive tools for visualizing mobile objects data, there are no capabilities for summarizing or aggregating these data, meaning that it is difficult to scale these applications to collections of trajectories without visual confusion. Shaw et al. (2008) addresses this issue by using clustering techniques to extract representational summary paths from trajectory collections.

2.2 Mobility Mining

The problem of extracting meaningful information from large databases is not unique to mobile objects data and GIS-based time geography. Knowledge Discovery from Databases (KDD) is the attempt to extract novel patterns hidden in massive digital databases through efficient computational techniques. The objective is to generate unexpected and interesting hypotheses that can be investigated further using standard inferential and confirmatory techniques. Geographic Knowledge Discovery (GKD) is a subset of KDD that attempts to discover novel spatio-temporal patterns in massive digital geographic datasets using scalable geocomputational techniques. GKD techniques exploit the unique characteristics of geographic data such as spatial dependency and heterogeneity. In addition, GKD tools can handle complex spatial properties such as the size and shape of geographic objects, and relationships among objects such as distance, direction and connectivity (Han et al. 2001; Miller and Han 2009). As the size and complexity of geospatial data increases, leveraging geocomputational techniques with geovisualization is essential to help manage the GKD process and interpret its results (Andrienko and Andrienko 2008).

Mobile objects data also creates unique challenges for the knowledge discovery process. Andrienko and Andrienko (2008) envision a specialized knowledge discovery process for these data called *mobility mining*. The mobility mining process involves three major steps:

1. *Trajectory reconstruction.* This involves processing the raw stream of mobility data to obtain the individual object trajectories. It also involves methods for efficient storage and access of these trajectories.
2. *Knowledge extraction.* This involves the application of spatio-temporal and mobile objects data mining methods to discover novel and useful information in these data. Possible patterns include *clusters* or groups of similar trajectories, *frequent patterns* reflecting repeatedly followed paths or subpaths and *classifications* based on behavioral rules extracted from the trajectories (also see Dodge et al. 2008).
3. *Knowledge delivery.* Extracted patterns are seldom direct knowledge; rather, these patterns must be evaluated based on their interestingness, interpreted relative to pertinent background knowledge and communicated in a manner appropriate for improving policy and decision-making in real-world applications.

Our main concern in this research is the rapid summarization of data as a first step in the knowledge extraction process. A well-known technique in online analytical processing (OLAP) technique is the *data cube*. The data cube is an operator that allows users to generate all possible cross-tabulations of the data at different levels of aggregation to provide synoptic summaries of the database (see Gray et al. 1997; Han and Kamber 2006). Shekhar et al. (2001) extended the data cube to the *map cube* that can handle the geographic components of the data and visualize them in concert with the cross-tabs and aggregations. The *traffic cube* is a further extension for handling spatio-temporal traffic data (see Lu et al. 2009; Shekhar et al. 2001, 2002). However, these methods require a fixed geography and spatial aggregation scheme and cannot be applied directly to trajectory summarization.

In addition to database aggregation methods, data visualization techniques enable simple and intuitive interactions of mobile object data and humans. The objective is to find interesting patterns, trends and relationships especially in mobile objects datasets, supporting knowledge construction about mobility behavior (Miller and Han 2009). Dynamic visual exploration is useful in understanding the structure of the dataset, raising questions about movement patterns, and facilitating identification of meaningful combinations of variables in further map representation and analysis (Wood and Dykes 2008). Interactive visualization methods associated with visualization software environment have been proposed to enhance the quality of pattern detection. For example, GeoTime is a three-dimensional visualization environment designed to visualize and analyze trajectories of mobile objects (Kapler and Wright 2005; Kraak 2003). Another widely used method for visual data exploration is trajectory aggregation. Data mining methods such as clustering support visual detection of clusters of mobile objects for data aggregation (Andrienko et al. 2009; Rinzivillo et al. 2008; Schreck et al. 2008). Although these studies successfully illustrate the importance of analysing spatio-temporal dynamics within a visualization environment, they heavily rely on locational and temporal information only. Few studies have explored the mobile object patterns from the attribute domain (Skupin 2008; Kraak and Huisman 2009).

An exploratory visualization technique designed specifically for mobile objects data is the *space-time cube* (Kraak 2003). The space-time cube visualizes spatio-temporal data in a three-dimensional environment that the user can manipulate through rotating, projecting, scaling and other visual browsing techniques (Kraak 2003) (Fig. 1). In addition, Leonardi et al. (2010) developed the *T-warehouse* for data warehouses system designed for trajectory data.

2.3 Data Aggregation and Similarity Measures

As noted above, a barrier to meaningful visualization of mobility databases is difficulty in extracting meaningful patterns from mobile objects data. *Data aggregation techniques* are methods for reducing the size of data to extract general patterns (Andrienko and Andrienko 2008). Several researchers have proposed time-based aggregation to summarize and analyze mobile objects data. For example, Hornsby and Egenhofer (2002) developed a framework that enables space-time

Fig. 1 The space-time cube
concept

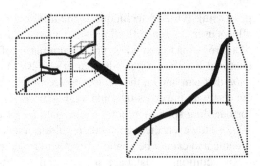

queries in multiple time granularities for space-time paths and prisms. In addition, there are some efforts to combine time geographic concepts and data summarization methods such as aggregation and clustering. These software tools visualize mobile object trajectories in two spatial dimensions and time, and provide capabilities to group trajectories based on location during a time period of interest (Pfoser and Theodoridis 2003; Shaw et al. 2008; Kapler and Wright 2005).

An emerging data aggregation technique for mobility data is *similarity measures*. Similarity measures can be used to analyze whether different mobile objects exhibit correspondence in terms of a metric distance function such as Euclidean distance (Sinha and Mark 2005; Yanagisawa et al. 2003), the Hausdorff distance measure for two point sets (Huttenlocher et al. 1993; Shao et al. 2010), the Frèchet distance for polygonal curve similarity (Eiter and Mannila 1994), and longest common subsequence (LCSS) for measuring similarity in time-series data (Vlachos et al. 2002). However, similarity measures alone do not allow the analyst to explore similarity at multiple scales (e.g., Laube et al. 2005). In addition, similarity measures can be computationally complex, although progress has been made with respect to scalable heuristics (Andrienko et al. 2009; Shao et al. 2010; Sinha and Mark 2005).

Another trend in similarity measure is trajectory descriptors. Trajectory descriptors are metrics of mobility physical characteristics such as location, direction and speed. These measures can serve as a basis for aggregating or grouping individual paths for summarization, improving the clarity of the visualization (Laube et al. 2005; Sinha and Mark 2005). Trajectory descriptors can be calculated at an individual sample location and can be extended into interval and/or global scales (Dodge et al. 2008). However, most studies focus on spatial and temporal domain; it is rare to examine dynamics within the attribute domain, that is, the evolution of non-locational properties over time such as the trajectory geometry and other physical movement parameters. This is also rarely linked with the growing area of geographic data mining and knowledge discovery (Skupin 2008; Skupin and Hagelman 2005).

The toolkit described in this paper combines similarity techniques with user-defined temporal granularity parameters to facilitate exploratory trajectory aggregation at varying levels of movement generality. Furthermore, the techniques developed in this visualization toolkit are computationally efficient and can be scaled to large databases and embedded in other exploratory techniques and processes. In addition, this paper describes a user-interactive visualization environment to summarize and explore mobile objects data based on the movement attributes of mobile object trajectories at different levels of temporal granularity. The next section of this paper discusses these time aggregation and similarity measure techniques, and the visualization environment that implements these techniques.

3 Methodology

3.1 Overview

This chapter develops time aggregation methods and similarity measures to enhance the discovery of multi-scale patterns in mobile objects data. There are several steps to analyze the mobile objects data within the interactive visualization tool proposed in this research (see Fig. 2). First, all the data are stored in a database in order to be extracted later as queries. Second, time aggregation methods allow the user to determine a time range of interest and temporal granularity within the selected time range to reconstruct individual trajectories at different levels of movement generality. Third, given these reconstructed trajectories, the user can apply similarity measures to aggregate individual trajectories based on location or attribute to aggregate multiple trajectories into synthetic trajectories that reflect collective movement patterns. This process can be repeated until the user of the toolkit finds meaningful patterns. We embed these techniques within a space-time cube environment that allow visual exploration and statistical summaries of the aggregated and grouped mobile objects data.

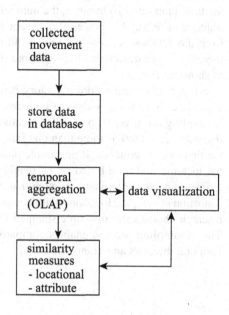

Fig. 2 Flowchart of the analysis

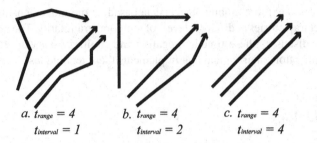

a. $t_{range} = 4$ *b.* $t_{range} = 4$ *c.* $t_{range} = 4$
    $t_{interval} = 1$ $t_{interval} = 2$ $t_{interval} = 4$

Fig. 3 Example of time range and time interval parameters

3.2 *Time Granularity and Trajectory Reconstruction*

Temporal granularity is a critical parameter for visual data exploration as well as data mining and statistical analysis since it can cause substantial difference in the results of visualization and analysis (Hornsby 2001). Visualization with coarse time granularity is more appropriate to explore broad scale movement while visualization with refined time granularity is more suitable for detailed movement of the mobile objects (Hornsby and Egenhofer 2002).

Two parameters for determining time granularity when reconstructing mobile object trajectories are the *time range* and *time interval*. Time range is the time period queried from the database. For example, if the user wants to visualize parts of trajectories at time between 10:00 and 11:00, '1 hour' is the time range. On the other hand, time interval is the granularity within the time range; the minimum time unit that divides time range equally. For example, if the time range is 1 hour and the time interval is 10 minutes, the number of time stamp is six. The reconstructed trajectories reflect the choices of range and interval. To illustrate, assume trajectories from the database such as the ones as illustrated by Fig. 3a. As the time interval increases, three trajectories become more similar as in Fig. 3b, and exactly the same as shown in Fig. 3c.

Since LATs often record trajectory data using independent sampling rates, we normalize the trajectory data to common sampling times using simple temporal resampling rules. We map recorded locations and times to the interval that includes that sample. If there is more than one sampled point within the interval, we choose the first one in sequence. If no sample point falls within an interval, we interpolate the location and time based on its neighboring intervals. This resampling rule is efficient and scalable; however, note that choosing longer time interval may cause distortion of sampled locations of trajectories because the algorithm proposed in this research chooses the first time-stamped location within the chosen time interval. This resampling process enables comparison of trajectories recorded at different temporal intervals and granularities.

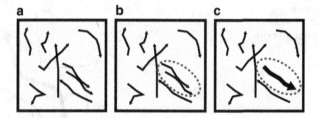

Fig. 4 Similarity aggregation: (**a**) individual trajectories; (**b**) detection of a cluster based on locational similarity; (**c**) summary trajectory

Fig. 5 Calculating locational similarity

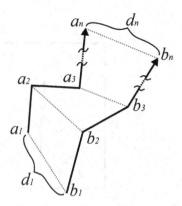

3.3 Similarity Measures

Locational Similarity

This function measures the similarity between trajectories based on their spatial footprints, allowing users to aggregate trajectories that are spatially proximal given the selected time interval and range. This measure is useful in finding where and when mobile objects are moving together in a sequence of time. Urban commuting behavior, normal crowd flow and animal flocking are example movement patterns that can exhibit locational similarity. Figure 4 illustrates the general process.

An efficient method to measure locational similarity is to calculate the Euclidean distance between two locations of mobile object trajectories at specified time intervals (Steiner et al. 2000). A Euclidean distance of zero indicates that two trajectories visit the same locations in space-time. Trajectories that share the same spatial locations but at different times will have a higher locational similarity score, as will trajectories that diverge in space, even if they share the same origin, destination or some intermediate locations (Fig. 5).

If trajectories have a high degree of locational similarity, we can meaningfully aggregate these into a trajectory that summarizes those locations. A simple and

Fig. 6 Aggregating
trajectories with locational
similarity by vector averaging

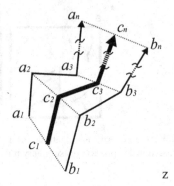

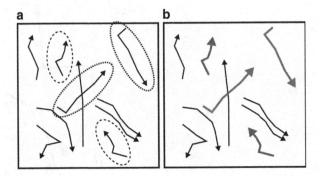

Fig. 7 Attribute similar trajectories: (**a**) detection; (**b**) identification

tractable method is to treat each polyline segment as a vector and finding the average of the corresponding vectors. Figure 6 illustrates the process for the two trajectory case for clarity.

While the Euclidean distance measure is straightforward, it suffers from sensitivity to outliers. Other possible distance measures include the Hausdorff and Fréchet distances. The Hausdorff distance is the maximum of the minimum distances between two curves; however, it can be misleading since it does not consider any temporal sequencing in the curves. The Fréchet distance captures the sequences within each curve (see Alt et al. 2003). We use the Euclidean distance measure for simplicity and scalability. However, our methods are not limited to Euclidean distances, and continuing development of the toolkit could include other distance measures for comparison purposes.

Attribute Similarity

In contrast to locational similarity, attribute similarity concerns intrinsic attribute properties of the trajectories regardless of their location or orientation in space. These measures can be used to categorize trajectories into groups based on similar attribute properties (see Fig. 7).

Fig. 8 Attribute similarity
between two space-time paths
based on three dimensions

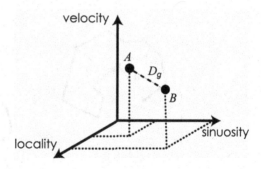

The attribute similarity functions in the visualization toolkit extract five measurable attributes of attribute similarity. These are *sinuosity, direction, velocity, locality* and *spatial range* (described in detail below). The indices map the trajectory to a point in a multidimensional space (see Wentz 2000). The Euclidean distance within this space represents the similarity or dissimilarity of trajectories based on attribute. Figure 8 illustrates this in three-dimensional space for clarity. $D_g = 0$ indicates that two trajectories are exactly the same in terms of attribute, with increasing positive values indicating greater attribute dissimilarity. The user sets a maximum D_g values as a threshold to detect groups of similar trajectories for aggregation.

Although reducing complex attribute similarity properties to a single point in multidimensional space results in information loss, it creates a simple measure for efficient clustering, as well as input into other data mining techniques. This places a burden on the user to explore a wide range of similarity thresholds and assess the resulting summary patterns. It is therefore critical that the implemented system have a high degree of user interactivity.

The subsections below describe the five attribute similarity indices. The five indices are not independent, nor do they exhaust all aspects of attribute similarity; see Andrienko et al. (2008), Dodge et al. (2008), and Huang et al. (2008) for discussions of other attribute similarity measures. The indices proposed in this research are semantically clear properties that can be captured in an efficient, scalable manner and expressible as dimensionless metrics for ease of comparison. The user can apply all the indices simultaneously, or any subset depending on the nature of the data and the relevant questions to be explored.

- *Sinuosity.* Sinuosity measures the deviation of the trajectory from a straight line. It is the ratio of the total length of the trajectory and the Euclidean distance between the origin and destination:

$$Sinuosity = \frac{d_E}{d_p} \tag{1}$$

where d_p is the total length of the trajectory and d_E is the Euclidean distance between the origin and the destination. This index varies between zero and one,

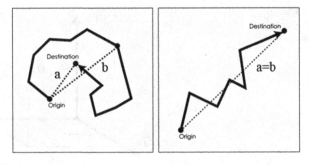

Fig. 9 Locality. Left side illustrates a low locality score; right side illustrates a high locality score

with one corresponding to a straight line and values closer to zero indicating a more sinuous path. The index is zero in the degenerate case of stationary behavior.

- *Direction.* This index captures the relative or egocentric direction of the trajectory:

$$Direction = \frac{\bar{D}}{180} \tag{2}$$

where $\bar{D}$ is the average egocentric direction of the line segments comprising the trajectory. Each line segment's egocentric direction is relative to the previous segment. Note the contrast with locational similarity: this considers directionality but from a cardinal perspective (e.g., two trajectories must travel in the same cardinal direction to have a low locational similarity value). In this index, two trajectories must have analogous tendencies with respect to turning directions to be similar. Since the egocentric direction ranges from $-180°$ (left hand side direction, or counter clockwise direction) to $180°$ (right hand side direction, or clockwise direction), the value of *Direction* ranges between -1 and $+1$.

- *Velocity.* Velocity indicates the relative speed of the object during the time period:

$$Velocity = \frac{\bar{V}}{V_{max}} \tag{3}$$

where $\bar{V}$ is the average velocity of a trajectory, and V_{max} is the maximum velocity in the data based on the used-defined temporal granularity parameters. Zero indicates stationary behaviour and one indicates matching the maximum velocity in the sample.

- *Locality.* Locality is the ratio between the distance between a trajectory's origin and its final destination and the distance between the origin and the farthest location in the trajectory (relative to the origin; see Fig. 9). This is a measure of the relative focus of the trajectory with respect to its initial and final location within the chosen temporal range.

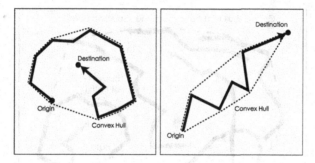

Fig. 10 Convex hull of a path

$$Locality = \frac{L_{OD}}{L_{OF}} \qquad (4)$$

where L_{OD} is the Euclidean distance between the origin and the destination, and L_{OF} is the distance between the origin and the farthest recorded location from the origin. With respect to human movement behavior, trajectories with higher locality scores (closer to one) tend to be focused and single purpose, while trajectories with lower locality scores (closer to zero) tend to be more leisurely and/or multipurpose. For example, with respect to animal movement, low locality scores may indicate searching or foraging behavior. Using the origin as a basis for calculating locality may appear arbitrary: one could choose the destination instead. However, using the origin as a reference reflects the common practice in transportation science of characterizing trips based on their origin. It is also possible to use the origin and destination as joint references for a more complete depiction, but this would make the behavior of the index less transparent.

- *Spatial range*. Spatial range measures the relative spatial extent of the movement. It is the area of convex hull that contains a trajectory divided by the area of the convex hull that contains all the trajectories in the database:

$$Spatial\, Range = \frac{A_{path}}{A_{all}} \qquad (5)$$

where A_{path} is the area of convex hull that contains individual trajectory (see Fig. 10) and A_{all} is the area of convex hull that contains all of the trajectories (see Fig. 11). A spatial range closer to zero indicates that the trajectory covers relatively little territory, while a spatial range closer to one indicates a more expansive territory for the object. A convex hull provides a relatively accurate measure of spatial range (relative to other measures such as the minimum bounding rectangle) with reasonable computational cost. The toolkit utilizes the Graham scan algorithm: this has the worse-case time complexity of $O(n \log n)$; this is better than quadratic and therefore scalable (Sedgewick 1990).

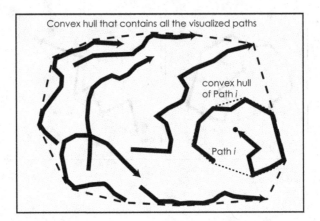

Fig. 11 Spatial range

Once the system calculates attribute similarity values, a next step is to generate groupings of trajectories based on these values. We look for natural groupings using an efficient density-based clustering method called DBSCAN (Ester et al. 1996). This is a spatial clustering method that looks for regions with sufficiently high density and forms clusters of arbitrary shapes. Although clustering methods designed explicitly for trajectory data are available (see Han et al. 2009), we chose DBSCAN due to its ability to incorporate the user-selected similarity indices discussed above as well as its scalability. DBSCAN also does not require designating the number of clusters *a priori*.

DBSCAN requires two input parameters: (i) an ε -*neighborhood* (expressed as a radius) for searching around each data point; (ii) *min-points* or the minimum number of data points required for in a neighborhood to be included. DBSCAN finds clusters by searching the ε -neighborhood of each data point, starting with an arbitrary point. If the ε -neighborhood of a point contains more than min-points, a new cluster is generated with that point designated as a *core object*. The algorithm iteratively adds data points that meet the search radius and density criteria to the core objects until no more points can be added. DBSCAN is efficient: it has the worst-case complexity of $O(n \log n)$ if a spatial index is used (Han and Kamber 2006; Han et al. 2009).

We customized DBSCAN in the following manner. First, the distances for finding neighboring points are the distances provided by the selected similarity functions. Second, we set min-points as one by default, indicating a cluster can be created from only two trajectories. Determining the ε -neighborhood for searching is more complex and can require trial-and-error exploration. To facilitate this, the visualization toolkit reports a set of statistical values for the data, including minimum and maximum values of the spatial coordinates in each dimension. A third modification is the inclusion of a maximum radius (*max-* ε) to limit the search around each point for scalability purposes. We set *max-* ε equal to twice the ε -neighborhood as a default, although this can be user-modified. This clustering algorithm is basically a combination of both DBSCAN and OPTICS (Han and Kamber 2006).

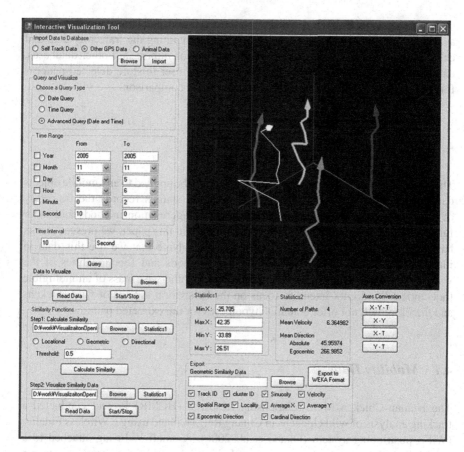

Fig. 12 Graphical user interface of the visualization tool

3.4   Visualization Toolkit

The visualization toolkit in this paper is based on the concept of the space-time cube (Kraak 2003). The toolkit encompasses three major functionalities: (i) individual trajectory reconstruction based on time granularity (time aggregation); (ii) trajectory aggregation and/or grouping based on trajectory similarity (similarity functions) and; (iii) data visualization, data summarization, and data export for further analysis. We developed the visualization toolkit using the C# programming language. In addition, Microsoft SQL Server 2005 provides the functionality for data storage and query support.

Figure 12 illustrates the main GUI. The user can visualize and explore these individual trajectories, aggregate or group trajectories based on their apparent similarity and extract summary statistical properties for the aggregated or clustered trajectories. The user can also visualize the trajectories in three dimensions (two-dimensional space and time), as well as project the trajectories into any two of the dimensions (*xy*, *xt* and *yt*). Arbitrary rotation of the visualization interface is also available. It is also possible to export data files containing the trajectory data and calculated parameters at any level of aggregation for import into other databases, data mining or statistical software. This allows the user to store detected patterns for further investigation, or as reference patterns for comparison.

4   Case Study

The aggregation methods proposed in this study can be applied to any mobile objects data where a sequence of time-stamped location stamps records the trajectory of each object. To illustrate the effectiveness of aggregation methods and other functionalities in the visualization toolkit proposed in this paper, we present results based on wild chicken tracking data in Thailand. Note that we use this data to illustrate the capabilities of the aggregation methods and functionalities of the visualization toolkit for extracting patterns from an otherwise mass of unintelligible trajectories. We do not intend to focus on the behavioral aspects of wild chickens and therefore will not offer potential hypotheses for the extracted patterns.

4.1   Mobility Data

The Human-Chicken Multi-relationship Research (HCMR) Project conducted a tracking analysis of wild chickens in Chiang Rai, Thailand using a Wireless Fidelity (WiFi) positioning system (Okabe et al. 2006). A small WiFi tag attached to a chicken's body records the location and time. The WiFi data tracking system consists of six devices, namely tags to stick to the chickens' legs, activator, receiver, Power over Ethernet (PoE), WiFi access point, and management engine. The weight of the tag is 35 g. Spatial resolution is 1 m and time resolution is 1 s. The fine time resolution of 1 s allows flexibility to analyze the data from detailed temporal scales to coarse temporal scales.

The study area is the 200 square-meter land under cultivation inside the Chiang Rai Livestock Research and Technology Transfer Center. Figure 13 illustrates the study setting, including the facilities and the locations of the chickens at one moment in time. There are eight concrete one-storied houses (H1 through H8) in the field: two of them are residential houses (H1 and H5) and the rest of them are empty houses. H2 is the preparation room for experimental appliances and H6 is the room for the data management engine that includes the software package that processes signals of location data sent via the WiFi access point and displays the locations of tags (Okabe et al. 2006). CH1 through CH3 are the locations of chicken houses. Since all the residents in two residential houses leave for agricultural work outside of

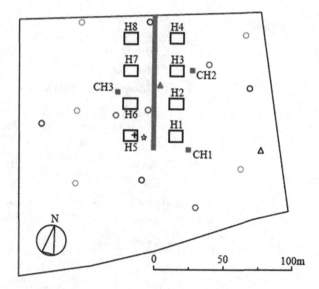

Fig. 13 Setting for wild chicken movement study (cited from Okabe et al. 2006)

the study field, chickens can move freely all over the study area. There are eighteen chickens in total (circle shape symbols in Fig. 13): there are three groups and each group consists of six chickens respectively. There are other symbols representing feeding sites (gray-colored triangle), trees (black, hollow triangle and star-shaped symbol), and the location of underfloor in the house H5 (cross-shaped symbol).

This research utilizes the movement data of 18 chickens with 2,979,359 time stamps from November 5th to November 8th 2005. Although the WiFi system uses x and y coordinates for locations, the location coordinates used in the system are independent from the geographic coordinates. The maximum spatial extent of the whole movement of the chickens can be expressed by maximum and minimum coordinates for both x and y coordinates. The minimum and maximum x coordinate are -83.79 and 80.39 respectively (164.18 in total for east–west extent), the minimum and maximum y coordinate is -53.71 and 49.42 (103.13 in north–south extent).

4.2 Toolkit Functionality

We now illustrate the toolkit functionality by showing results from querying the database at different levels of temporal granularity and aggregating the trajectories based on similarity at the specified granularity.

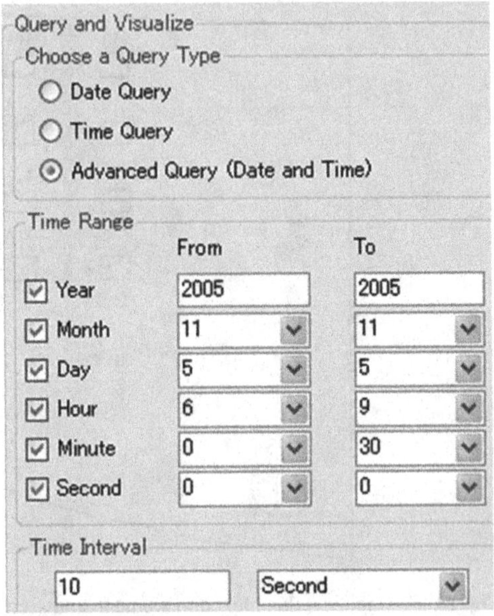

Fig. 14 Temporal
aggregation using OLAP in
the toolkit

Query and Visualize

Choose a Query Type

○ Date Query

○ Time Query

◉ Advanced Query (Date and Time)

Time Range

	From	To
☑ Year	2005	2005
☑ Month	11	11
☑ Day	5	5
☑ Hour	6	9
☑ Minute	0	30
☑ Second	0	0

Time Interval

10   Second

Time Granularity and Trajectory Reconstruction

The proposed toolkit in this research enables the user to specify the time range and time interval of interest. Figure 14 shows the temporal aggregation GUI illustrating an OLAP query. The user first chooses the type of query from three options; date query, time query, or the advanced query that can specify both date and time. The user chooses the query type based on one's interest to extract a portion of trajectory data. The second parameter is the time range. This example is the case when the time range is between 6:00:00 and 9:30:00 on November 5th in 2005. The user can specify each temporal resolution from the drop down menus. The third and the last parameter is the time interval. This example shows the case of 10 s. The user again can choose the time resolution from the drop down menu. The options are years, days, hours, minutes, and seconds.

Figures 15 and 16 show the effects of time granularity on trajectory reconstruction. Figure 15 illustrates the reconstructed trajectory collection for the wild chicken data at three different time ranges on November 5th, 2005, with the time interval provided by the data (1 s). Figure 16 illustrates the reconstructed wild chicken trajectory collection at three time intervals for a fixed time range from 6:00 to 17:00 on November 5th, 2005. As Fig. 15 suggests, it is difficult to extract distinct patterns at the highest level of temporal granularity. Even with a relatively low time range (6:00–9:00), the trajectory collection is an undistinguished mass. This problem becomes more acute as the time range increases. Note that the map at the bottom of the visualization window shows that spatial extent of the whole movement also expands as the time range increases. Figure 16 indicates that changing the time

6:00 - 9:00	6:00 - 12:00	6:00 - 17:00

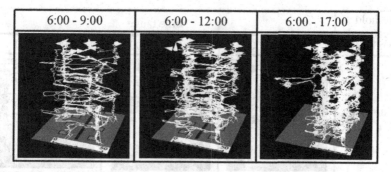

Fig. 15 Reconstructed trajectories at different time ranges

5 second	1 minute	10 minute

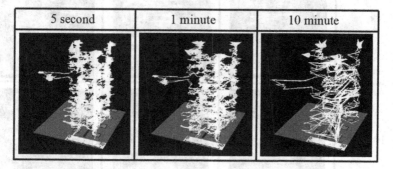

Fig. 16 Reconstructed trajectories at different time intervals

interval can mitigate this problem to a substantial degree: the trajectories are more generalized and patterns are more easily discernable as the time intervals become coarser. The visualization toolkit allows the user to visualize the trajectory collection at the time range of interest and interactively change the time interval until an appropriate granularity level is achieved for the data and questions at hand.

Locational Similarity

After the user has selected the time range and interval, the toolkit allows aggregation of similar trajectories to detect clearer patterns from the data. Figure 17 illustrates the effects of locational similarity-based trajectory on the visualized patterns at different time ranges for the wild chicken data in a three dimensional view. Figure 17 compares the unaggregated trajectories from Fig. 15 (top row in Fig. 15) with aggregated trajectories based on a strict locational similarity threshold of 5.0 (middle row) and a relaxed locational similarity threshold of 10.0 (bottom row). The top row once again illustrates the problem with unaggregated trajectories: it is difficult to discern any generalized patterns. In contrast, aggregation based

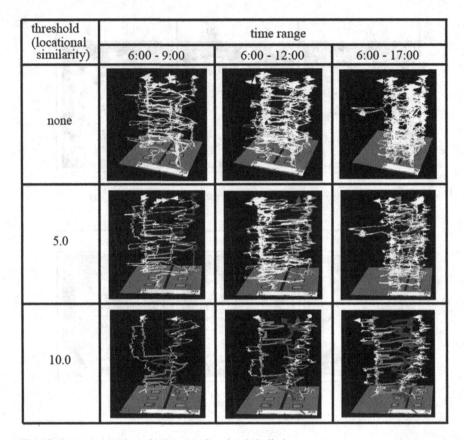

threshold (locational similarity)	time range		
	6:00 - 9:00	6:00 - 12:00	6:00 - 17:00
none			
5.0			
10.0			

Fig. 17 Aggregated trajectories based on locational similarity

on locational similarity facilitates the detection of movement patterns. Summary paths (rendered blue, green and red) were extracted at both threshold levels. At the strict locational similarity threshold of 5.0, three summary paths were extracted for the time range 6:00–9:00, but only two paths for the longer time ranges of 6:00–12:00 and 6:00–17:00. In addition, these summary paths are occluded by the outliers (rendered in white) meaning that they represent a relatively small number of the sample trajectories. Three summary paths that are relatively stable across all three time ranges were extracted at the more relaxed threshold of 10.0. Also, these summary paths are easier to discern since the number of outliers is smaller. The appropriate value for this threshold must be determined by user-interactivity: the toolkit facilitates this process.

The toolkit also reports statistical data for the aggregated paths: this can help with user interpretation of the results. Table 1 provides some of the statistical data for the aggregate trajectories in Fig. 17. Based on the statistics in Table 1, directional values become close to zero as the time range increases in all three clusters regardless of

Table 1 Statistical information of summary trajectories in Fig. 17

Locational similarity threshold	Cluster color	Time range	Length	Mean velocity	Mean cardinal direction	Mean egocentric direction
5.0	Red	6:00–9:00	34.00	2.62	15.25	178.28
		6:00–12:00	475.00	3.44	1.19	8.63
		6:00–17:00	1273.25	2.91	0.42	2.22
	Green	6:00–9:00	520.45	1.45	1.38	17.48
		6:00–12:00	894.50	1.24	0.55	−144.40
		6:00–17:00	1388.52	1.05	0.54	1.53
	Blue[a]	6:00–9:00	422.89	0.78	1.11	−127.30
10.0	Red	6:00–9:00	34.00	2.62	15.25	178.28
		6:00–12:00	475.00	3.44	1.19	8.63
		6:00–17:00	1250.28	2.76	0.56	1.04
	Green	6:00–9:00	485.73	1.35	1.78	7.72
		6:00–12:00	945.98	1.31	0.04	7.96
		6:00–17:00	1658.74	1.26	0.31	6.83
	Blue	6:00–9:00	532.38	0.99	1.46	−20.03
		6:00–12:00	939.12	0.87	0.74	−26.49
		6:00–17:00	1706.96	0.86	0.10	−7.99

[a]No clusters occurred during the 6:00–12:00 and 6:00–17:00 time ranges

the locational similarity threshold. This implies the movements are not fixed in a certain direction. In addition, the mean cardinal direction is always close to zero degree in all cases, indicating that movement tend to direct all directions. Also, the mean velocity of red cluster increases as the time range increases from 6:00–9:00 to 6:00–12:00 while the mean velocity of other clusters decrease as the time range increases regardless of the locational similarity threshold.

To help users identify the numbers of detected clusters, it is useful to observe the change in the number of detected clusters in relation to time interval and time range. Figure 18 traces relationship between the change in the number of detected clusters of wild chicken data in three different time intervals and locational similarity threshold values. As is shown in the Fig. 18, number of clusters changes in a similar manner in all three time intervals indicating there are similar clusters detected regardless of the difference in time intervals. Figure 19 shows the relationship between the change in the number of detected clusters of wild chicken data in three different time ranges and locational similarity threshold values. The number of clusters varies with different time ranges although the trend in change of the number of clusters is similar in all three time ranges. Generally, the higher the locational similarity threshold, the more trajectories are likely to be included in fewer numbers of trajectories, resulting in detecting only one cluster when the threshold value is very large. However in this case, the number of clusters converges to either three or four indicating there are distinct differences in those three or four groups of movement.

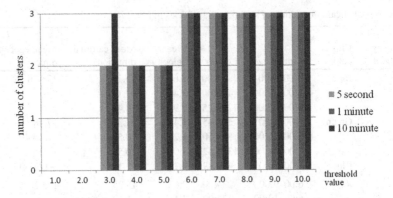

Fig. 18 Change in the number of detected clusters of wild chicken data by locational similarity measure with respect to time interval

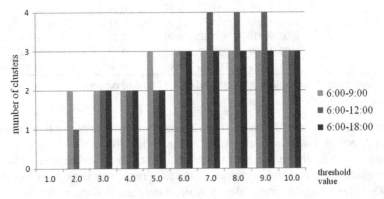

Fig. 19 Change in the number of detected clusters of wild chicken data by locational similarity measure with respect to time range

In addition to the three dimensional view, Fig. 20 clearly shows the spatial distribution of the movements of chickens. The aggregated trajectories with a relaxed locational similarity threshold of 10.0 detected the same three clusters from shorter time range (6:00–9:00) to the longest time range (6:00–17:00). This indicates three important findings. First, the locational similarity measure successfully detected three clusters that are reported in Okabe et al. (2006) that used the same data set. Okabe et al. (2006) reported that there are three main groups of chickens that behave as flocks for the entire study period. Second, the locational similarity measure detected the clusters at similar locations throughout the day, which implies the consistency in the movement of three groups of chickens. Chickens in this case tend to move as groups although there are some outliers (trajectories in white). Third, locations of all three detected clusters overlap or are close to the location where the food is (the triangle point in Fig. 14). The chickens did not move through wide areas of the study area but stayed close to where houses are located.

threshold (locational similarity)	time range		
	6:00 - 9:00	6:00 - 12:00	6:00 - 17:00
none			
5.0			
10.0			

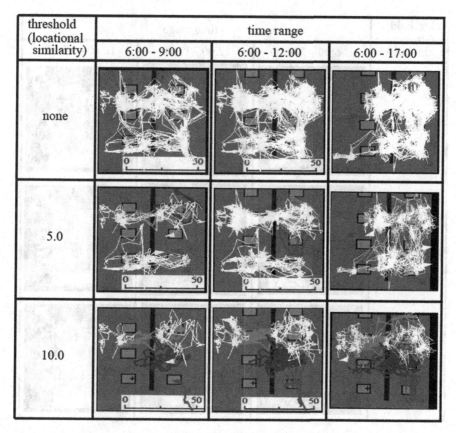

Fig. 20 Two dimensional view of aggregated trajectories based on locational similarity

Attribute Similarity

Figure 21 shows visualizations of the wild chicken trajectory collection at different attribute similarity thresholds and for different time ranges. The time interval is fixed at 10 s and all five attribute similarity functions are invoked. Comparable to locational similarity, a strict attribute similarity threshold (0.1) only detects a small number of trajectories for clustering, but a more generous threshold (0.5) identifies a greater number of candidates. Obvious patterns appear when attribute similarity function is applied compared with the visualization without any similarity functions (top row). In addition, similar patterns with respect to attributes tend to appear at similar locations: paths with similar attribute properties tend to occur in proximity, suggested coordinated movement behavior. It is also interesting that the locations of clusters detected in two time range, 12:00–13:00 and 16:00–17:00, are similar to each other. Chickens may move the same areas in different time ranges: this suggests repetitive movement patterns. Okabe et al. (2006) also reported that there are some

threshold (attribute similarity)	time range		
	6:00 - 7:00	12:00 - 13:00	16:00 - 17:00
none			
0.1			
0.5			

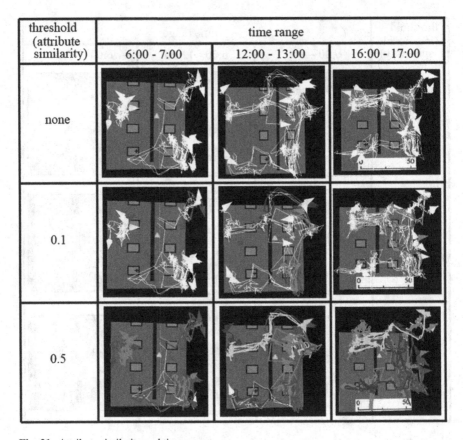

Fig. 21 Attribute similarity and time ranges

chickens that follow other chickens throughout a day. These are mainly hens that follow a cock that leads his own chicken group. The attribute similarity function may have detected this type of flocking behavior.

5   Discussion and Conclusion

This research develops an interactive visualization toolkit based on temporal granularity and spatial similarity to explore and discover multi-scale mobility patterns in mobile objects databases. The toolkit facilitates highly interactive visual exploration of mobile trajectories at varying levels of temporal granularity and thresholds for trajectory aggregation based on locational and attribute similarity among paths at the specified granularity level. A case study of wild chicken mobility dataset shows that combination of both time granularity and trajectory aggregation facilitates mobility pattern detection. The interactive temporal aggregation method with OLAP in the proposed toolkit is the first step to explore trajectory data to mitigate the difficulty of exploring complex movement patterns. In addition, visualization with similarity thresholds provides distinct views of the movement data, with some discovered patterns being robust across different granularities and others being dependent on these parameters. The flexibility of the temporal querying and trajectory aggregation also allows the discovery of temporally recurrent mobility patterns: both locational and attribute similarity measures detected mobility patterns that Okabe et al. (2006) also uncovered with the same dataset.

There are some remaining challenges for continued development of the mobility visualization toolkit. Although the similarity functions in this research are scalable and effective at detecting similar mobility patterns, there are many ways to assess trajectory similarity. Other ways of measuring locational and attribute similarity should be explored, as well as other definitions of path similarity distinct from the two dimensions explored in this research. One possible way is to examine the properties that can be extracted from trajectories. There are other characteristics that can be calculated from trajectories other than five characteristics described as attribute similarity such as average travel distance, average x coordinate location and y coordinate location of a trajectory, and so on. Behavioral characteristics such as number of activities, number of visiting locations activity duration time are also candidates.

The trajectory summarization methods in the toolkit consider only the central tendency (mean values) of trajectory parameters such as velocity and direction. Searching for patterns based on central tendencies is reasonable for exploratory analysis since these patterns should be tested using confirmatory techniques before being accepted as knowledge. Nevertheless, a more complete representation of trajectory similarity would consider the dispersion (variance) of these parameters. A research frontier is to incorporate parameter variance in the summarization methods in a manner that is both scalable and intuitive to the analyst.

A related research challenge is linking the scalable, exploratory tools in this research to confirmatory techniques. The patterns discovered using the visualization methods are only hypotheses: these should be tested using more powerful analytical and statistical methods. These tools could be used in conjunction with the techniques in this toolkit to confirm and further analyze the tentative patterns discovered through visual exploration.

References

Alt H, Knauer H, Wenk C (2003) Comparison of distance measures for planar curves. Algorithmica 38:45–58

Andrienko G, Andrienko N (2008) Spatio-temporal aggregation for visual analysis of movements. In: IEEE symposium on Visual Analytics Science and Technology (VAST), Columbus, 21–23 Oct 2008, pp 51–58

Andrienko N, Andrienko G, Pelekis N, Spaccapietra S (2008) Basic concepts of movement data. In: Giannotti F, Pedreschi D (eds) Mobility, data mining and privacy: geographic knowledge discovery. Springer, Berlin, pp 15–38

Andrienko G, Andrienko N, Rinzivillo S, Nanni M, Pedreschi D, Giannotti F (2009) Interactive visual clustering of large collections of trajectories. In: Proceedings of the IEEE symposium on Visual Analytics Science and Technology (VAST), IEEE Computer Society Press, pp 3–10

Brimicombe A, Li Y (2006) Mobile space-time envelopes for location-based services. Trans GIS 10:5–23

Dodge S, Weibel R, Lautenschütz AK (2008) Towards a taxonomy of movement patterns. Inform Vis 7:240–252

Eiter T, Mannila H (1994) Computing discrete Fréchet distance. Technical University of Wien, CD-TR 94/64

Ester M, Kriegel HP, Sander J, Xu X (1996) A density-based algorithm for discovering clusters in large spatial databases. In: Proceedings of 2nd international conference on knowledge discovery and data mining, Portland, pp 226–231

Gray J, Chaudhuri S, Bosworth A, Layman A, Reichart D, Venkatrao M, Pellow F, Pirahesh H (1997) Data cube: a relational aggregation operator generalizing group-by, cross-tab and subtotals. Data Mini Know Disc 1:29–53

Hagerstrand T (1970) What about people in regional science? Papers Regional Sci Assoc 24:7–21

Han J, Kamber M (2006) Data mining – concepts and techniques. Morgan Kaufmann, San Francisco

Han J, Kamber M, Tung AKH (2001) Spatial clustering methods in data mining. In: Miller HJ, Han J (eds) Geographic data mining and knowledge discovery, 2nd edn. Taylor and Francis, London, pp 74–109

Han J, Lee JG, Kamber M (2009) An overview of clustering methods in geographic data analysis. In: Miller HJ, Han J (eds) Geographic data mining and knowledge discovery, 2nd edn. Taylor and Francis, London, pp 149–187

Hornsby KS (2001) Temporal zooming. Trans GIS 5:255–272

Hornsby KS, Cole S (2007) Modeling moving geospatial objects from an event-based perspective. Trans GIS 11(4):555–573

Hornsby KS, Egenhofer M (2002) Modeling moving objects over multiple granularities, special issue on spatial and temporal granularity. Ann Math Artif Intel 36:177–194

Huang Y, Chen C, Dong P (2008) Modeling herds and their evolvements from trajectory data. In: Proceedings of the 5th international conference on geographic information science (GIScience 2008), Park City, pp 90–105

Huttenlocher DP, Klanderman GA, Rucklidge WJ (1993) Comparing images using the Hausdorff distance. IEEE Trans Pattern Anal Mach Intell 15(9):850–863

Kapler T, Wright W (2005) GeoTime information visualization. Inform Vis 4:136–146

Kraak M-J (2003) The space-time cube revisited from a geovisualization perspective. In: Proceedings of the 21th international cartographic conference, Durban, pp 1988–1995

Kraak MJ, Huisman O (2009) Beyond exploratory visualization of space-time paths. In: Miller HJ, Han J (eds) Geographic data mining and knowledge discovery. CRC, Boca Raton, pp 431–443

Kwan M-P (2000) Interactive geovisualization of activity-travel patterns using 3D geographical information systems: a methodological exploration with a large data set. Transport Res Part C 8:185–203

Langran G (1992) Time in geographic information systems. Taylor and Francis, London

Laube P, Imfeld S, Weibel R (2005) Discovering relative motion patterns in groups of moving point objects. Int J Geogr Inform Sci 19:639–668

Leonardi L, Marketos G, Frentzos E, Giatrakos N, Orlando S, Pelekis N, Raffaeta A, Roncato A, Silvestri C, Theodoridis Y (2010) T-warehouse: visual olap analysis on trajectory data. In: IEEE 26th international conference on data engineering (ICDE), 2010, Long Beach, pp 1141–1144. IEEE

Li C, Longley P (2006) A test environment for location-based services applications. Trans GIS 10:43–61

Lu C-T, Boedihardjo AP, Shekhar S (2009) Analysis of spatial data with map cubes: highway traffic data. In: Miller HJ, Han J (eds) Geographic data mining and knowledge discovery, 2nd edn. Taylor and Francis, London, pp 223–245

Miller HJ (1991) Modelling accessibility using space-time prism concepts within geographical information systems. Int J Geogr Inform Syst 5:287–301

Miller HJ (2005) A measurement theory for time geography. Geogr Anal 37:17–45

Miller HJ, Han J (2009) Geographic data mining and knowledge discovery: an overview. In: Miller HJ, Han J (eds) Geographic data mining and knowledge discovery, 2nd edn. Taylor and Francis, London, pp 3–32

Okabe A, Satoh T, Okabe K, Imamura E, Morathop S, Jailanka C, Ratanasermpong S, Hayashi Y, Akishinonomiya F (2006) Acquisition of spatio-temporal data of free-range chicken using a Wireless Fidelity (WiFi) positioning system and spatial analysis of the obtained data. The papers and proceedings of the geographic information systems association, vol 15, pp 395–400

Peuquet DJ, Duan N (1995) An event-based spatiotemporal data model (ESTDM) for temporal analysis of geographical data. Int J Geogr Inform Syst 9:7–24

Pfoser D, Theodoridis Y (2003) Generating semantics-based trajectories of moving objects. Comput Environ Urban Syst 27:243–263

Rinzivillo S, Pedreschi D, Nanni M, Giannotti F, Andrienko N, Andrienko G (2008) Visually driven analysis of movement data by progressive clustering. Inf Vis 7(3-4):225–239

Roberti M (2003) Wal-Mart spells out RFID vision. [online]. RFID Journal. http://www.rfidjournal.com/. Accessed 16 June 2003

Schreck T, Bernard J, Tekusova T (2008) Visual cluster analysis of trajectory data with interactive Kohonen Maps. In: Proceedings of the IEEE symposium on visual analysis science and technology (VAST 2008), Columbus, pp 3–10

Sedgewick R (1990) Finding the convex hull. In: Sedgewick R (ed) Algorithms in C. Addison-Wesley Publishing Company, Reading, pp 359–372

Shao F, Cai S, Gu J (2010) A modified Hausdorff distance based algorithm for 2- dimensional spatial trajectory matching. In: The 5th international conference on computer science & education, 24–27 Aug, Hefei, pp 166–172

Shaw S-L, Yu H, Bombom LS (2008) A space-time GIS approach to exploring large individual-based spatiotemporal datasets. Trans GIS 12:425–441

Shekhar S, Lu CT, Tan X, Chawla S, Vatsavai RR (2001) A visualization tool for spatial data warehouses. In: Miller HJ, Han J (eds) Geographic data mining and knowledge discovery. Taylor and Francis, London, pp 74–109

Shekhar S, Lu CT, Liu R, Zhou C (2002) CubeView: a system for traffic data visualization. Intelligent transportation systems. In: Proceedings of the fifth IEEE international conference on intelligent transportation systems, pp 674–678

Sinha G, Mark DM (2005) Measuring similarity between geospatial lifelines in studies of environmental health. J Geogr Syst 7:115–136

Sistla P, Wolfson O, Chamberlain S, Dao S (1998) Querying the uncertain position of moving objects. In: Etzion O, Jajodia S, Sripada S (eds) Temporal database: research and practice. Springer, New York, pp 310–337

Skupin A (2008) Visualizing human movement in attribute space. In: Agarwal P, Skupin A (eds) Self-organizing maps: applications ain geographic information science. Wiley, Chichester

Skupin A, Hagelman R (2005) Visualizing demographic trajectories with self-organizing maps. GeoInformatica 9(2):159–179

Steiner I, Bürgi C, Werffeli S, Dell'Omo G, Valenti P, Tröster G, Wolfer DP, Lipp H-P (2000) A GPS logger and software for analysis of homing in pigeons and small mammals. Physiol Behav 71:589–596

Turchin P (1998) Quantitative analysis of movement: measuring and modeling population redistribution in animals and plants. Sinauer Associates, Sunderland

Vlachos M, Kollios G, Gunopulos D (2002) Discovering similar multidimensional trajectories. In: Proceedings of the 18th international conference on data engineering, 2002, San Francisco, pp 673–684

Wentz EA (2000) A shape definition for geographic applications based on edge, elongation, and perforation. Geogr Anal 32–1:95–112

Wentz EA, Campell AF, Houston R (2003) A comparison of two methods to create tracks of moving objects: linear weighted distance and constrained random walk. Int J Geogr Inf Sci 17:623–645

Wood J, Dykes J (2008) Spatially ordered treemaps. IEEE Trans Vis Comput Graph (Proceedings Visualization/Information Visualization 2008), 14(6):1348–1355

Worboys M, Duckham M (2004) GIS: a computing perspective, 2nd edn. Taylor and Francis, London

Yanagisawa Y, Akahani JI, Satoh T (2003) Shape-based similarity query for trajectory of mobile objects. In: Chen M-S et al (eds) Mobile data management, LNCS 2574. Springer, Berlin, pp 63–77

Yu H, Shaw S-L (2008) Exploring potential human activities in physical and virtual spaces: a spatio-temporal GIS approach. Int J Geogr Inf Sci 22:409–430

Yuan M (2001) Representing geographic information to support queries about life and motion of socio-economic units. In: Frank A, Raper J, Cheylan J-P (eds) Life and motion of socio-economic units. Taylor and Francis, London, pp 217–234

About the Authors

Rodger A. Brown is a research meteorologist with the NOAA National Severe Storms Laboratory in Norman, Oklahoma, and is an adjunct professor of meteorology at the University of Oklahoma. To help improve National Weather Service warnings, he uses Doppler weather radar to study the evolution of tornadoes and other hazardous weather phenomena produced by severe thunderstorms. He also participates in a data mining project that looks for precursor signatures within numerically modeled severe thunderstorms that can lead to tornado production and thereby further increase the lead time of tornado warnings.

Jörg Dallmeyer received his bachelor degree (2008) as well as master degree (2009) in computer science from the Goethe University of Frankfurt am Main, Germany. Since December 2009, he has been a PhD student at the chair for Information Systems and Simulation at Goethe University Frankfurt. His research interests are actor-based simulation for the field of traffic simulation under consideration of multimodal traffic and the building of simulation systems from geographical information.

G. Cervone et al. (eds.), *Data Mining for Geoinformatics: Methods and Applications*,
DOI 10.1007/978-1-4614-7669-6, © Springer Science+Business Media New York 2014

P. Franzese received a PhD in aerospace engineering from the Polytechnic University of Turin in 1995 and was a postdoctoral fellow at CSIRO Division of Atmospheric Research in Aspendale, Australia, in 1996–1998. Formerly a research associate professor at George Mason University since 1996, he is currently with the environmental consulting company Ecology and Environment, Inc.

Sandra Geisler is a research assistant and PhD student at the Information Systems chair (http://www.dbis.rwth-aachen.de) at the RWTH Aachen University headed by Professor Dr. Matthias Jarke. She completed her undergraduate studies at the TU Dortmund and received her diploma in computer science from the RWTH Aachen University in 2008. She wrote her diploma thesis while working at Philips Technologie GmbH, Europe, in Aachen, which resulted in the publication of two patents. Currently, she is working in the research initiative UMIC (Ultra-Highspeed Mobile Information and Communication, http://www.umic.rwth-aachen. de) on mobile data management applications. Furthermore, she is involved in the project "Cooperative Cars eXtended" where she is investigating data stream-based solutions for traffic management applications using C2X communication. She is a co-organizer of the International Workshop on Information Management for Mobile Applications series held in conjunction with VLDB. She is also guest editor of the Elsevier journal *Pervasive and Mobile Computing* for the special issue on "Information Management in Mobile Applications."

Hajime Imura received his PhD in computer science in 2011 from the Graduate School of Information Science and Technology, Hokkaido University, Sapporo, Japan. He did his postdoc at the Meme Media Laboratory of Hokkaido University and is now working there as a specially appointed assistant professor. His research interests include document image retrieval, information retrieval, machine learning, and data mining.

Tetsuo Kobayashi is assistant professor of geography at the Florida State University in Tallahassee, Florida, USA. His research interest is mobility data analysis in geographic information science. Specifically, he has been working on the development of data mining and visualization techniques for mobile objects data in various applications, such as human mobility in urban systems, transportation analysis, and tactics of team sports, especially soccer. He is also working on urban sustainability with the focus on spatiotemporal analysis of green buildings.

Micke Nicander Kuwahara is a specially appointed assistant professor at the Meme Media Laboratory of Hokkaido University in Sapporo, Japan. He holds two bachelor degrees, one in computer science from the Royal Institute of Technology, Stockholm, Sweden, and another in education from the Stockholm University, Sweden. During the last 5 years, he has been the lead architect and system developer of the new generation of IntelligentPad-inspired Meme Media implementation called the Webble World.

Andreas D. Lattner received his diploma (2000) and doctoral degree (2007) in computer science from the University of Bremen, Germany, and the venia legendi for his professorial thesis in computer science from the University of Trier, Germany (2012). From 2000 to 2007, he was working as a research scientist in the Intelligent Systems department of the Center for Computing Technologies (TZI) at the University of Bremen. Since 2007, he has been working as a postdoctoral researcher at the chair for Information Systems and Simulation at Goethe University Frankfurt. His research interests include knowledge discovery in simulation experiments, temporal pattern mining, and multiagent systems.

Amy McGovern is an associate professor in the School of Computer Science at the University of Oklahoma and an adjunct associate professor in the School of Meteorology. Her research focuses on developing spatiotemporal machine learning and data mining models, primarily for severe weather applications. She received her PhD (2002) and MS (1998) in computer science from the University of Massachusetts, Amherst, and her BS (1996, honors) from Carnegie Mellon University.

Harvey J. Miller is professor of geography at the University of Utah in Salt Lake City, Utah, USA. His research and teaching focus on the application of geographic information science and spatial analytic techniques to study things moving in geographic space – in particular, humans within cities. He is also interested in individuals' use of transportation and communication technologies to generate mobility and accessibility, and the implications for sustainable transportation systems, livable communities, and public health. In addition to approximately 100 scientific publications in peer-refereed journals and edited books, Harvey is author (with Shih-Lung Shaw) of *Geographic Information Systems for Transportation: Principles and Applications* (Oxford University Press) and editor (with Jiawei Han) of *Geographic Data Mining and Knowledge Discovery*, second edition (CRC Press).

Pavel Moiseets graduated from Far Eastern State University, Russia, in 2007 (Faculty of Applied Programming and Mathematics) and worked at the Pacific Institute of Geography previously. He is currently working toward his PhD degree in computer science in the Knowledge Media Laboratory at Hokkaido University. His research interests include GIS, information visualization, databases, data mining, and smart object federation.

Christoph Quix is an assistant professor at the chair of Information Systems (Informatik 5) of RWTH Aachen University, Germany, where he also received his PhD degree in computer science in 2003. His research focuses on metadata management, data integration, and semantic web technologies. He published about 60 publications in scientific journals and international conferences. He was member of the program committee of several conferences in the area of databases and data modeling (e.g., ER, ICDE, and ODBASE).

Ron Resmini is an associate professor in the College of Science at George Mason University, Fairfax, Virginia, and a research scientist in the Advanced ISR Solutions Department of the MITRE Corporation, McLean, Virginia. He specializes in visible to infrared multi- and hyperspectral imagery (HSI) remote sensing, the geological and geophysical sciences, and the analysis, design, and development of algorithms for processing and analysis of remotely sensed information. His current research activities focus on the design, development, implementation, and testing of algorithms for resolved and unresolved (subpixel) target detection, classification, identification, and characterization in hyperspectral data with emphases on linear and nonlinear spectral mixing models and statistical signal processing techniques. His other research interests include mathematical modeling of natural processes observed in remotely sensed data, modeling of spectral signatures of natural and man-made materials, the characterization of HSI data in hyperspace, and the utilization and evaluation of radiative transfer models as applied to the spectral remote sensing of the Earth's surface. With over 19 years of industrial and academic experience in HSI remote sensing, Ron Resmini has supported numerous government programs demonstrating and advancing the utility of remotely sensed spectral information for a wide range of applications.

Derek H. Rosendahl is a doctoral student in the School of Meteorology at the University of Oklahoma (OU). His interests include climate change and variability, paleoclimate, global and regional climate modeling, earth system science, severe convective storms and storm-scale modeling, and the impacts of climate change on human and natural systems. He is also interested in the science-policy interface, having been selected to attend the American Meteorological Society's 2007 Summer Policy Colloquium in Washington, DC. Derek received his BS and MS in meteorology from OU with his thesis research on identifying precursors to strong low-level rotation within an ensemble of numerically simulated supercell thunderstorms using a data mining approach. His doctoral research involves using a multi-thousand member global climate model ensemble to assess uncertainties in global and North American regional climate change projections. Upon receiving his PhD, Derek will become a postdoctoral researcher at the Department of Interior South Central Climate Science Center, where he will conduct research on uncertainties in climate change projections across North America with a focus on South Central USA.

Mark Salvador has worked with military sensors and systems since 1990. His experience spans hardware and software development for programs in Intelligence, Surveillance, and Reconnaissance (ISR); missile defense; tactical weapon systems; research and development; and field operations. He has to his credit numerous papers and presentations in the area of hyperspectral remote sensing and has developed several automated processing systems and algorithms for hyperspectral analysis. In 2011, Dr. Salvador received the Director of National Intelligence Science and Technology Team Award for his work in support of Operation Enduring Freedom. Dr. Salvador is chief architect and develops advanced hyperspectral systems at Exelis Inc.

Jonas Sjöbergh received his PhD in computer science (on natural language processing) in 2006 from the Royal Institute of Technology (KTH) in Stockholm, Sweden. He did a postdoc at the Language Media Lab of Hokkaido University and is now working as a specially appointed assistant professor at the Meme Media Lab of Hokkaido University. His research interests include natural language processing, artificial intelligence, machine humor, smart object federation, data mining, and information visualization.

Yuzuru Tanaka is a full professor of knowledge media architecture at the Department of Computer Science, Graduate School of Information Science and Technology, Hokkaido University, and the director of Meme Media Laboratory, Hokkaido University. He is also an adjunct professor of the National Institute of Informatics. His recent research areas cover meme media architectures, knowledge federation frameworks, and proximity-based federation of smart objects and their application to digital libraries, e-Science, clinical trials, and social cyber-physical systems for the optimization or improvement of social system services, such as snow plowing and snow removal in Sapporo, Japan, and urban disaster management and response.

Ingo J. Timm received his diploma (1997), PhD (2004), and venia legendi (professorial thesis) (2006) in computer science from the University of Bremen. From 1998 to 2006, he had been a PhD student, research assistant, visiting and senior researcher, and managing director at the University of Bremen, Technical University Ilmenau, and Indiana University – Purdue University Indianapolis (IUPUI). In 2006, Ingo Timm was appointed full professor for Information Systems and Simulation at Goethe University Frankfurt. Since fall 2010, he holds a chair for Business Informatics at the University of Trier. Ingo Timm's research focuses works on information systems and knowledge-based systems in logistics and medicine. His special interests lie in the strategic management of autonomous software systems, actor-based (multiagent-based) simulation, and knowledge-based support to simulation systems.

Tetsuya Yoshida received his Dr.Eng. degree from the University of Tokyo, Japan, in 1997. He was an assistant professor at Osaka University in Japan from 1997 to 2004 and works as associate professor at Hokkaido University in Japan since 2004. His research interests include machine learning, data mining, and social network analysis.

About the Editors

Dr. Guido Cervone is associate professor of geoinformatics in the Department of Geography and Institute for CyberScience at the Pennsylvania State University. He is also affiliate scientist with the Research Application Laboratory (RAL) at the National Center of Atmospheric Research (NCAR). His research expertise is in machine learning and geoinformatics, and his main interest is the mining of spatial and temporal remote sensing, model and social media big data associated with natural, man-made, and technological hazards. He worked on the theoretical development and implementation of symbolic and evolutionary machine learning systems. He developed a new methodology based on non-Darwinian evolution to identify the source characteristics of an unknown toxic atmospheric release.

He sits on the advisory committee of the United Nation Environmental Programme (UNEP), Division of Disasters and Early Warning Assessment (DEWA). His research is currently being funded by the Department of Transportation and by the Office of Naval Research.

Dr. Nigel Waters is a professor in the Department of Geography and Geoinformation Science and director of the Geographic Information Science Center of Excellence at George Mason University. His present research involves the use of

G. Cervone et al. (eds.), *Data Mining for Geoinformatics: Methods and Applications*, DOI 10.1007/978-1-4614-7669-6, © Springer Science+Business Media New York 2014

GIS techniques and social media data for transportation and health research and is supported by the US Department of Transportation and the National Institutes of Health. He is the editor of *Cartographica: The International Journal for Geographic Information and Geovisualization*, which is published by the University of Toronto Press. He is a member of the Board of Directors of the University Consortium for Geographic Information Science. He was the 2010 Henrietta Harvey Distinguished Lecturer, at Memorial University, Newfoundland.

Dr. Jessica Lin is an associate professor in the Department of Computer Science at George Mason University (GMU). She received her PhD degree from the University of California, Riverside, in June 2005. Her research interests encompass broad areas of data mining, especially data mining for large temporal and spatiotemporal databases, text, and images. Over the years, she has collaborated with researchers from various domains including medicine, earth sciences, manufacturing, national defense, and astronomy. Her research is partially funded by NSF, US Army, and Intel Corporation. Dr. Lin has been member of the program committee of many international conferences in the area of data mining. She teaches advanced topics on data mining at GMU, concentrating on mining multimedia and high-dimensional data.

Printed in the United States
By Bookmasters